HISTORY OF THE ZODIAC

ROBERT POWELL

HISTORY OF THE ZODIAC

SOPHIA ACADEMIC PRESS

SAN RAFAEL, CALIFORNIA

First published in the USA
by Sophia Academic Press
© Robert Powell 2007

No part of this book may be reproduced
or transmitted, in any form or by
any means, without permission

For information, address:
Sophia Academic Press, P.O. Box 151011
San Rafael, California 94915, USA

Library of Congress Cataloging-in-Publication Data

Powell, Robert, 1947–
History of the zodiac / Robert A. Powell.—1st ed.

p. cm.

Includes bibliographical references.
ISBN 1 59731 152 9 (pbk: alk. paper)
ISBN 1 59731 153 7 (hardcover: alk. paper)
1. Zodiac—History. 2. Astronomy, Ancient. I. Title
BF1726.P677 2007
133.5'3—dc22   2006100661

CONTENTS

ACKNOWLEDGEMENTS 1

INTRODUCTION 3

1 SIGNS AND CONSTELLATIONS OF THE ZODIAC 11

Divisions in Space and Time 13
The Historical Approach 16
Ptolemy and Hipparchus 18
Euctemon's Solar Calendar 20
The Progression of the Apsides 27

2 THE PRECESSION OF THE EQUINOXES 30

The Phenomenon of Precession 32
The Rate of Precession: Historical Survey 35
Hipparchus' Discovery of the Precession of the Equinoxes 37
The Alleged Babylonian Discovery of Precession 39
Precession in Relation to the Sidereal Zodiac 44

3 THE TROPICAL ZODIAC 57

The Tropical Zodiac: A Universal Solar Calendar 62
The History of the Tropical Zodiac 64
Further Historical Remarks 74
The Tropical Zodiac in Astrology 79
The Substitution of the Tropical Zodiac for the Sidereal Zodiac 80

4 THE SIDEREAL ZODIAC IN EGYPT 84

The Decans 87
The Decans in Relation to the Zodiac 91

5 THE SIDEREAL ZODIAC IN MESOPOTAMIA 96

Normal Stars 97
Normal Stars in Relation to Signs of the Zodiac 100
Historical Remarks 109

6 THE SIDEREAL ZODIAC IN INDIA 114

The Nakshatras (Lunar Mansions) 114
The Redefinition of the Nakshatras 121
The Introduction of the Zodiac into India 124
Later Indian Astronomy 126
The Indian Astronomical Tradition of the Great Year 128
How the Indian Sidereal Zodiac differs from the Babylonian 131
A New Definition of the Indian Sidereal Zodiac 136

APPENDIX I: BABYLONIAN STAR CATALOGUE RECONSTRUCTED 140

The Intrinsic Definition of the Babylonian Zodiac 140
Babylonian Star Catalogue 141

APPENDIX II: NEWTON'S CHRONOLOGY 195

AFTERWORD 203

BIBLIOGRAPHY 208

ACKNOWLEDGMENTS

ISAAC NEWTON REMARKED in a letter of 5th February 1676: "If I have seen further it is by standing on the shoulders of Giants."[1] In quoting Newton the author of this thesis on the definition of the zodiac certainly does not intend to say that he has seen further than other historians of astronomy, but rather he wishes to acknowledge that it is thanks to "standing on the shoulders of giants" that this thesis was able to be written. Three names of such "giants" of the twentieth century are: Otto Neugebauer, David Pingree, and B.L. van der Waerden. In turn, their research was made possible through the efforts of many others. The initial decipherment of astronomical cuneiform texts began with the work of the Jesuit Fathers Johann Strassmaier, who transcribed numerous texts in the British Museum, and Joseph Epping, who undertook the task of working out the astronomical and mathematical content of these texts. Following in the footsteps of Epping, Father Franz Xaver Kugler revolutionized the study of Babylonian astronomy. Kugler's works, beginning with *Die Babylonische Mondrechnung* (1900), uncovered an entire lost world of mathematical astronomy of a high level of sophistication. Further important contributions to the recovery of Babylonian astronomy were made during the twentieth century by Asger Aaboe, John Britton, Peter Huber, Hermann Hunger, Erica Reiner, Francesca Rochberg, Abraham Sachs, Johann Schaumberger, Noel M. Swerdlow, Christopher Walker, Ernst Weidner, and others.

The author acknowledges his indebtedness to all these scholars of the history of astronomy, and especially to Professors Otto Neugebauer (1899–1990) and David Pingree (1933–2005) of Brown University, Rhode Island, without whose scholarship this thesis would not have been possible. Professor Pingree's research on the cross-cultural transmission of science has paved the way for a new and deeper understanding as to how the zodiac, as one of the most significant innovations of Babylonian astronomy, was transmitted to other cultures. Moreover, together with Assyriologists Erica Reiner and Hermann Hunger, he has accomplished for early Babylonian astronomy—in terms of advancing our knowledge and understanding of it—what Otto Neugebauer achieved for later Babylonian astronomy. Professor Neugebauer's contribution to the history of

---

1. I.B. Cohen, "Isaac Newton," *Dictionary of Scientific Biography* 10, p55.

ancient astronomy—covering Babylonian, Egyptian, Greek astronomy, and other areas—is unparalleled. Having read an earlier draft of this thesis on the definition of the Babylonian zodiac and the subsequent history of the zodiac, Professor Neugebauer remarked that "such a history would fill a gap in the literature."[1] Likewise Professor B.L. van der Waerden (1903–1996) of the University of Zürich read an earlier draft of this thesis on the definition of the zodiac. Commenting upon the central question of this thesis concerning the stars Aldebaran and Antares defining the fiducial axis of the Babylonian zodiac, he remarked: "I fully agree with your hypothesis, namely: Taurus 15° = Aldebaran and Scorpio 15° = Antares by definition."[2] B.L. van der Waerden's great accomplishment was the mapping out of the stages of the birth of astronomy and mathematics. The author wishes to express his gratitude to Professor B.L. van der Waerden for his helpful comments concerning this thesis on the history of the zodiac, and also to Professor Konrad Rudnički of Jagiellonian University, Cracow, for his encouragement. Lastly, the author acknowledges his debt of gratitude to Peter Treadgold (1943–2005) for providing the computations underlying both the reconstructed Babylonian star catalogue and Table 7 (The Fifty Brightest Zodiacal Stars) tabulated in Appendix I and to Dianna Marsden for her help in typing this thesis.

1. Letter of May 18, 1977.
2. Letter of March 30, 1983.

INTRODUCTION

WRITING THIS BOOK—my Ph.D. thesis—has been a thirty-year odyssey. It started in 1974 when I began doing the initial research on the history of the zodiac at the libraries of the British Museum and the Warburg Institute in London. After three years, during which time I wrote the first draft, it became apparent that the scope of my thesis went beyond the range of interest of my supervisor. Right at that time, early 1978, while a guest at the Institute for Mathematics and Physics in Dornach, near Basel, Switzerland, I met Konrad Rudnicki, professor of astronomy at Jagiellonian University in Cracow, Poland, where Copernicus had studied. Through Professor Rudnicki as my supervisor I enrolled for completion of my Ph.D. at Jagiellonian University. However, when Poland's communist ruler, General Jaruzelski, declared martial law in Poland on December 13, 1981, disbanding the Solidarity trade union and placing its leader, Lech Walesa, under house arrest, I was advised by a well-meaning friend to discontinue the Ph.D. process at Jagiellonian University. After much deliberation, on November 2, 1982 I wrote a letter to Professor Rudnicki expressing my decision to withdraw from the doctoral procedure.

Some years went by. Then came the peaceful revolution that led to the collapse of communism in Poland and the other Eastern European countries. I now felt that it would be possible to resume the doctoral procedure at Jagiellonian University. Upon expressing this intention, Professor Rudnicki communicated that the basic requirements applying to the doctoral procedure at Jagiellonian University had changed completely. Now, he said, I would have to fulfill residency requirements, which was not possible for me (given my life circumstances). Further years went by. Then he learnt that it would be possible for me to complete the doctoral procedure at the Polish Academy of Sciences (Institute for the History of Science). This meant an intensive time of re-writing, i.e. updating my thesis, and then submitting the revised thesis to the Institute for the History of Science in Warsaw.

By way of introduction, therefore, this book on the history of the zodiac was submitted as a Ph.D. thesis to the Polish Academy of Sciences (Institute for the History of Science) with the title *The Definition of the Babylonian Zodiac and the Influence of Babylonian Astronomy on the Subsequent Defining of the Zodiac*. After taking the necessary exams and publicly defending my thesis in Warsaw on December 20, 2004, I was awarded a Ph.D. for my original contribution to the history of science in the sphere of the history of astronomy. This book is a revised and updated version of the Ph.D. thesis.

A natural question to ask is why the title *The Definition of the Babylonian Zodiac* was chosen rather than, for example, *The Conception of the Babylonian Zodiac* or *The Construction of the Babylonian Zodiac*? As outlined in this book, what I call the *intrinsic definition* of the Babylonian zodiac is referred to as the axis connecting two first magnitude stars, Aldebaran and Antares. The word *intrinsic* signifies that no reference to this defining axis of the Babylonian zodiac is to be found anywhere in the available cuneiform sources, and it is in this sense that the word *intrinsic* is used. (It is not emphasized in any way as a polarity to *extrinsic*.)[1] Further, the word *definition* is used to indicate that the axis in question was the basis for defining the longitudes of stars belonging to the star catalogue underlying the Babylonian zodiac in the sense that the other stars in this catalogue were assigned longitudes through measuring their distances from Aldebaran and Antares. Therefore it is the *defining* axis of measurement (fiducial axis)[2] of the Babylonian zodiac. Thus, while it is true that the Babylonian zodiac is both a conception and a construction, the axis defining this zodiac is the *defining axis*, i.e. it defines how the Babylonians conceived of their zodiac and how they constructed it by means of specifying a star catalogue by determining the longitudes of stars through measurement and observation, taking the defining axis as their point of departure, as clarified further in the summary at the end of this book.

Another natural question that may arise concerns the *scientific* nature of Babylonian astronomy. How scientific was it?

Van der Waerden's book *Science Awakening, Volume II: The Birth of Astronomy* offers the insight that it was pre-eminently the Babylonians who gave birth to the science of astronomy, which was then developed further by the Greeks. "Astronomy is the oldest physical science. It was highly developed by the Babylonians and the Greeks."[3] However, the Babylonians did not develop astronomy beyond the stage of applying arithmetical methods to it, whereas the Greeks were the first to apply geometry and trigonometry—including spherical trigonometry—to astronomy.

We now know that the Greeks borrowed certain astronomical parameters from the Babylonians.[4] Alone the use of the sexagesimal system (that is, in the

---

1. The polarity *intrinsic-extrinsic* is articulated in the articles by D. Lewis and P. Valentine (see Bibliography).
2. Fiducial = 'serving as a reliable or trustworthy basis of reckoning'.
3. Van der Waerden, SA II, p1.
4. Toomer (1988); cf. also Neugebauer, HAMA I, pp 279ff, 304, 544 and II, pp 591 ff.

measuring of time and also in the division of the circle into 360 degrees, each degree containing 60 minutes) attests Babylonian influence upon Greek and later developments of astronomy, including modern astronomy. We take the sexagesimal system (60-based system) for granted and tend to forget that it was thanks to the Babylonians that the sexagesimal system came about long ago, in the second millennium BC. Apart from the sexagesimal system the other major contribution to astronomy is the zodiac, a division of the ecliptic into twelve signs, each thirty degrees in length, that originated with the Babylonians in the fifth century BC as a coordinate system for specifying the location of stars and planets.

It is the origin of this coordinate system and its transmission from Babylon to other cultures which comprises the central content of this book. As indicated in the summary at the end, the introduction of the ecliptic coordinate system of the zodiac signified a major step forward in the development of astronomy as a science, for it meant that mathematical methods could be applied to the astronomical data collected through observation. It is in this sense—along the lines of van der Waerden's expression "science awakening: the birth of astronomy"—that the scientific nature of Babylonian astronomy, particularly after the introduction of the ecliptic coordinate system of the zodiac, is emphasized, i.e. with the Babylonians we witness the initial emergence of astronomy as a science. At the same time it is clear that in the case of Babylonian astronomy—in comparison with later Greek astronomy—it is a matter of a rather primitive level of science, without any critical self-reflection upon accuracy and procedures such as we find with the Greeks.[1]

A third natural question concerning the subject matter of this book is: Why is the focus of attention limited to Babylonian astronomy and its influence—through originating the zodiac—upon Greek, Egyptian (Hellenistic) and Indian (Hindu) astronomy? Or to put it another way: Why has Chinese astronomy or Meso-American astronomy or any other astronomical tradition not been taken into consideration? The simple answer is that, interesting though the astronomy of other cultures is, it would lead too far to explore them. The astronomical conceptions of Babylonia, Greece, Egypt and India are each major topics in themselves—particularly in relation to the first two of these four cultures—and a line had to be drawn somewhere. So this work on the history of the zodiac is limited to these four cultures (and a natural overspill to Roman culture in so far as the latter was the recipient of astronomy and astrology transmitted from Babylonia, Egypt and Greece). And, moreover, in focusing upon the zodiac, the fact is that the zodiac originated in

---

1. Rochberg (2002).

6 HISTORY OF THE ZODIAC

Babylonia as a sidereal ecliptic coordinate system that became transmitted to modern astronomy via Ptolemy in its metamorphosed Greek form (the tropical ecliptic coordinate system), which justifies limiting the focus of attention primarily to Babylonian and Greek astronomy, including the other cultures only in so far as they also are relevant to the theme of the history of the zodiac.

❦

There is no comprehensive work on the history of the zodiac available at the present time.[1] This thesis addresses the question of the original definition of the zodiac and its transmission to other cultures, dealing solely with the scientific aspect of the various definitions of the zodiac, without entering into the complexities of familiar zodiacal iconography. The time of origin of the first scientific definition is now known: "The zodiac of twelve signs of equal length ... was invented in Babylonia in the fifth century BC."[2] It is evident that the representations of the zodiac known to us from the Greek tradition of astronomy and astrology were in turn adopted from the Babylonian tradition.[3] The Babylonian origin of the zodiac is also apparent from the fact that many of the Greek names for the constellations are translations or modifications of the corresponding Babylonian names.[4] The earliest mention of these names is from Babylonian "astrolabes" (star lists arranged in a circular or rectangular pattern) thought to have been compiled around −1100,[5] where the 'Bull of Heaven', the 'Great Twins', the 'Scales', and the 'Scorpion' are among the

---

1. Van der Waerden, 'History of the Zodiac' (1953), gives an introductory overview concerning this theme, and Brack-Bernsen-Hunger, 'The Babylonian Zodiac: Speculations on its inventional significance' (1999), offers some thoughts concerning the observational practice of the Babylonians in relation to the zodiac. Cf. also Gleadow (1968).

2. Hunger-Pingree (1999), p17.

3. The indebtedness of the Greeks to the Babylonians is acknowledged in the middle of the fifth century BC by Herodotus, *Histories* (II: 109, 3); cf. also Toomer (1988) and Evans (1998), pp39–40.

4. Van der Waerden, SA II, p288.

5. Historians use the BC notation whereas astronomers generally use the minus (-) notation for dates in the pre-Christian era, since it is easier to compute time intervals using the minus notation. Historians write: 3 BC, 2 BC, 1 BC, AD 1, AD 2, AD 3, etc., whereas astronomers write this sequence: -2, -1, 0, +1, +2, +3, thus equating 1 BC with the year zero for the sake of computational consistency. This means that the historical year n+1 BC = −n astronomically. For example, 1101 BC = −1100. In this book, often the astronomical notation is used for dates belonging to the pre-Christian era, but—according to the context—the historical BC notation is also used frequently.

constellations listed.[1] A full list of names for zodiacal constellations appears for the first time in the MUL.APIN series of tablets inscribed in –686.[2] This was probably derived from earlier versions dating back to –1000 or earlier. Representations of some zodiacal constellations appear on boundary stones from the reign of Nebuchadnezzar I (–1123 to –1102).[3] By the time of the Persian period (–537 to –330) the Babylonian counterparts of the familiar Greek figures of zodiacal iconography were well established. In a text from the years –474 to –456, the division of the zodiacal constellations into twelve 30° sectors appears for the first time.[4] It is the scientific definition of this division—the original signs of the zodiac—that is central to this thesis, and also the transmission of the zodiac to other cultures.

The main contribution of this thesis is to answer the question as to *how* the zodiac of twelve signs of equal length was originally defined by the Babylonians. Among various conjectures and attempts to arrive at an understanding of the original definition, two analyses of Babylonian data with a view to discovering how the Babylonian zodiac was defined are noteworthy: in 1958 by Huber[5] and more recently by Kollerstrom.[6] The analysis by Huber was based exclusively on Babylonian data, whereas that by Kollerstrom was based on a small sample of Babylonian data together with a large amount of data from Greek astrological sources. As referred to in Chapter 5 of this thesis, Kollerstrom arrived at more or less the same result as Huber, but with a small difference of one degree in the definition of the zodiac. Both authors agree that the Babylonians utilized a sidereal zodiac, i.e. a zodiac whose twelve signs, each 30° long, are defined in relation to the fixed stars comprising the zodiac. The planetary longitudes given in terms of the Babylonian sidereal zodiac thus differ from corresponding longitudes given in terms of the tropical zodiac that is used in modern astronomy as a coordinate system.

> Babylonian (sidereal) longitudes may accordingly be compared against modern computed (tropical) longitudes by means of a correction factor that takes into account the constant of precession and the date of the data to be compared. The correction factor is that determined by P. Huber as a mean difference between ancient and modern longitudes, that $\Delta\lambda$ being 4°28' for the year –100.[7] Therefore, λ Babylonian = λ tropical + $\Delta\lambda$ where $\Delta\lambda$ = 3.08° + 1.3825° x (year date

---

1. Evans (1998), pp 5–11; cf. also Hunger-Pingree (1999), pp 50–57.
2. Hunger-Pingree (1989).
3. Van der Waerden, SA II, p 126.
4. Aaboe-Sachs (1969); cf. also Hunger-Pingree (1999), pp 184–185.
5. Huber (1958).
6. Kollerstrom (1997, 2001).
7. Huber (1958).

number). 3.08° is the correction factor for the year 0 and 1.3825° is the constant of precession per 100 years.[1]

This formula suggested by J.P. Britton[2] on the basis of Huber's analysis of Babylonian data enables the conversion from Babylonian sidereal longitudes to modern tropical longitudes and *vice versa*. However, it does not tell us *how* the Babylonian sidereal zodiac was defined. By its very definition as a sidereal zodiac, it must have been defined in relation to zodiacal fixed stars. But which stars were the defining (fiducial) stars? And how were these fiducial stars specified in relation to the signs and degrees of the zodiac?

Nothing has been found in the available cuneiform sources that would indicate an answer to these questions. The point of departure in this thesis—in exploring how the Babylonians defined the original zodiac—is to take seriously the statements of some Greek astrologers who were the recipients of Babylonian astronomy and astrology. These statements are examined in Chapter 5 of this thesis. Summarizing these statements briefly: the first magnitude star Aldebaran (α Tauri) is located at 15° in the sign of Taurus and the first magnitude star Antares (α Scorpii) is located at 15° in the sign of Scorpio.[3]

The main hypothesis of this thesis is that this location of Aldebaran and Antares in the middle of their respective signs constitutes the *intrinsic definition* of the Babylonian sidereal zodiac—intrinsic, because although it is nowhere explicitly stated in the cuneiform sources available to us, it is nevertheless the foundation upon which the zodiac rests. The content of this thesis supports the hypothesis that the Aldebaran-Antares axis was indeed the fiducial axis defining the Babylonian zodiac. On the basis of this conclusion the star catalogue underlying the Babylonian sidereal zodiac is reconstructed in an expanded form (following Ptolemy's star catalogue in the *Almagest*) in Appendix I, signifying an original contribution to the history of ancient astronomy by reconstructing the world's most ancient star catalogue, thus providing a standard ancient star catalogue against which astronomical data retrieved from antiquity may be compared. The fruit of this thesis, tabulated in Appendix I, is the reconstructed star catalogue underlying the Babylonian

---

1. Rochberg (1999), p57. On the basis of the definition of the Babylonian zodiac presented in this thesis (Appendix I), the value for $\Delta\lambda$ stands in need of slight modification. 1.3825° in 100 years corresponds to a rate of precession of 1° in 72.337 years. However, 1° in 72 years is a more accurate value, corresponding to 1.38889° per 100 years. Since $\Delta\lambda = 0$ in AD 220 (Appendix I), the correction factor for the year 0 amounts to 3.05556° (rather than 3.08°). [2.2 x 1.388889 = 3.05556]. Thus, modified $\Delta\lambda = 3.05556° + 1.38889$ x (year date number). In the year −100, modified $\Delta\lambda = 3.05556 + 1.38889 = 4.44445° = 4°27'$, closely agreeing with Huber's value of 4°28' for the year −100.
2. Ibid.
3. Neugebauer, HAMA II, p960 and Neugebauer-van Hoesen, *Greek Horoscopes*, p187.

sidereal zodiac. Three points should be noted with respect to this star catalogue:

(1) It is in a far more comprehensive form than the original star catalogue of the Babylonians. A fragment of the original—the world's first star catalogue—has been found. This fragment lists seven stars, only three of which are identified with their given longitudes.[1] These seven stars belong to the standard list of 32 reference stars (called *Normal Stars*) used by the Babylonians prior to (and also after) the innovation of the zodiac.[2] In reconstructing the Babylonian star catalogue it would suffice to tabulate the list of 32 Normal Stars with their ecliptic coordinates, whereas the reconstructed star catalogue in Appendix I tabulates the 1022 stars listed in Ptolemy's catalogue from the *Almagest*. The reason for this more expanded form, as indicated above, is to provide a comprehensive 'Babylonian style' star catalogue that can be used for the sake of comparison with ancient astronomical data from cuneiform or other sources.

(2) The original star catalogue of the Babylonians listed stars in sidereal ecliptic longitudes to the nearest whole degree, whereas the reconstructed Babylonian star catalogue in Appendix I—for reasons of accuracy—lists their sidereal ecliptic longitudes to degree and minute based on modern astronomical computations. If this star catalogue is to be useful to other researchers, it needs to be accurate. As indicated in Appendix I, the three stars retrieved *with their longitudes* from the original star catalogue are found to be accurately located in the reconstructed star catalogue. Rounding the longitudes of these three stars to the nearest degree yields precisely the longitudes given for these stars in the original star catalogue.

(3) The original Babylonian star catalogue listed stars only in terms of ecliptic longitude, without any indication of latitude. The latitudes of stars are given in the reconstructed Babylonian star catalogue since the latitudes could be of value for the sake of comparison with astronomical data retrieved from cuneiform or other ancient sources. Thus in recording the positions of the planets in relation to the set of 32 reference stars (Normal Stars) the Babylonians often indicated whether a planet was "above" or 'below' a star, which evidently signified an appraisal of the planet's latitude.[3] Similarly, the observation as to whether a planet was

---

1. Sachs (1952, 2); cf. also Hunger-Pingree (1999), p150.
2. Hunger-Pingree (1999), pp148–149.
3. Graßhoff (1999), p140.

'in front of' or 'behind' a star was evidently related to the planet's longitude.[1]

(4) In the original star catalogue of the Babylonians there was no indication of apparent stellar magnitude. For the sake of completeness apparent stellar magnitude is listed in the reconstructed star catalogue.

(5) The reconstructed Babylonian star catalogue is based on Ptolemy's star catalogue, the most famous and most comprehensive ancient star catalogue. However, there is a fundamental difference between the construction of the Babylonian star catalogue and that of Ptolemy. Like the Babylonians, Ptolemy used the ecliptic coordinate system of the zodiac—the system that originated with the Babylonians. However, whereas the Babylonian star catalogue uses *sidereal* ecliptic longitudes, Ptolemy's star catalogue utilizes *tropical* ecliptic longitudes of stars, i.e. measured from the vernal point defined as ♈ 0°. That is, Ptolemy borrowed the Babylonian innovation of the zodiac as his coordinate system but essentially redefined it. This thesis explores the original definition underlying the system of sidereal ecliptic longitudes of the Babylonians and also how it came about that the Babylonian zodiac became redefined by Greek astronomers.

This thesis thus explores the influence of Babylonian astronomy on the subsequent defining of the zodiac. After outlining basic considerations in the first two chapters, Chapter 3 is devoted to the history of the tropical zodiac in which by definition the zero point (♈ 0°) is equated with the vernal point. In relation to the tropical zodiac, a possible line of descent from Babylonian astronomy is considered in Chapter 3, where it is indicated that the ancient solar calendar from the MUL.APIN series of tablets[2] may have been the prototype of the tropical zodiac adopted by Greek astronomers. The tracing of this line of descent from Babylonian astronomy leading to the specification of the tropical zodiac is an original contribution of this thesis. Further, the transmission of the Babylonian zodiac to Greek, Egyptian (Hellenistic), and Indian (Hindu) cultures is explored in various chapters of this thesis, uncovering the indebtedness of all these cultures to Babylonian astronomy—in particular with regard to the original definition of the zodiac by the Babylonians.

---

1. Ibid.
2. Hunger-Pingree, p 61.

1

SIGNS AND CONSTELLATIONS OF THE ZODIAC

AT THE PRESENT TIME the word zodiac signifies for most people a division of the year into twelve months as follows:

Table 1

ZODIACAL SIGNS	CORRESPONDING MONTHS	NUMBER OF DAYS
Aries	21 March – 20 April	30 days
Taurus	20 April – 21 May	31 days
Gemini	21 May – 21 June	31 days
Cancer	21 June – 23 July	32 days
Leo	23 July – 23 August	31 days
Virgo	23 August – 23 September	31 days
Libra	23 September – 23 October	31 days
Scorpio	23 October – 22 November	30 days
Sagittarius	22 November – 22 December	30 days
Capricorn	22 December – 20 January	29 days
Aquarius	20 January – 19 February	30 days
Pisces	19 February – 21 March	30 days

The above dates for the twelve zodiacal months are for a 'typical' year, the dates listed may vary by a day or so from year to year.

In the general usage of the term 'sign of the zodiac', the sign of Aries is associated with the month extending from 21st March to 20th April; and it is said that during this period of time the sun is 'in' the sign of Aries.

But what does this mean to an astronomer? Suppose that an astronomer is able to carry out an observation of the sun on the day of the vernal equinox on the 20th or 21st March, when day and night are of equal length all over the world. On this day of the year, according to general understanding, the sun is just entering the sign of Aries. However, at the present time the position of the sun *astronomically* on the day of the vernal equinox is close to the tail of the southwestern Fish of Pisces (Figure 1). Hence the astronomer locates the sun in the constellation of Pisces on 21st March. In so doing he differs from those who refer the sun to the beginning of the sign of Aries on 21st March. There is a discrepancy between the general understanding of the movement of the sun and the specialist knowledge of the astronomer. According to the latter, the sun traverses the zodiacal constellations during the course of the year not through twelve equal zodiacal months (as given in Table 1), but through twelve *unequal* time periods (Table 2).

Figure 1

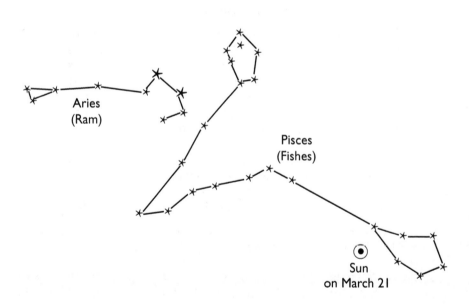

Table 2

ZODIACAL CONSTELLATION	DATES OF THE SUN'S PASSAGE THROUGH EACH CONSTELLATION
Ram	18 April—14 May
Bull	14 May—21 June
Twins	21 June—20 July
Crab	20 July—10 August
Lion	10 August—16 September
Virgin	16 September—31 October
Balance	31 October—23 November
Scorpion	23 November—18 December
Archer	18 December—19 January
Goat	19 January—16 February
Waterman	16 February—12 March
Fishes	12 March—18 April

The above dates are for a 'typical' year; the dates given may vary by a day or so from year to year.

DIVISIONS IN SPACE[1]—AND TIME

How may this difference in conception of the zodiac—the difference between the understanding of the general public and the astronomers—be reconciled? As discussed in detail in Chapter 3, the *tropical zodiac* originated as a division of the year into twelve months, commencing with the month of Aries starting at the vernal equinox. This division of the year comprises twelve solar months. On the other hand the *astronomical zodiac*, a division of the fixed stars into twelve unequal constellations, forms the background of the sun's passage through the stars in the course of one year. The term zodiac is applied to both these divisions. In the first case—that of the tropical zodiac—the sun passes through the twelve signs of the zodiac, spending one month in each sign during the course of the year. It originated in the fifth century BC. The months were named after the signs or constellations of the zodiac. The originator of this calendar scheme was Euctemon of Athens (432 BC).[2]

In the second case astronomers refer to the progression of the sun in space, through the twelve constellations of the astronomical zodiac, again in the

---

1. The term 'space' is used here in this section solely to emphasize the contrast with time. Rather than 'space', the proper astronomical term is 'celestial sphere'.
2. Pritchett-van der Waerden (1961).

course of one year. It is clear that on the one hand a division into invisible signs, originally conceived of as monthly time divisions, is referred to, while on the other hand the specified frame of reference is visible—namely the twelve zodiacal constellations. The invisible signs were initially defined as divisions in time, while the astronomical constellations are and always have been visible divisions in space. Accordingly it follows that originally the signs of the tropical zodiac were *temporal* divisions (of the year into twelve solar months) in relation to nature's yearly cycle, and that the constellations are *spatial* divisions (of the fixed stars into twelve unequal sectors) forming the background to the movement of the sun, moon, and planets. Thus the term *zodiac* is applied correspondingly in two quite distinct ways.

Both of these specifications of the zodiac—spatial and temporal—have a practical relevance. The astronomical zodiac, which divides the belt of fixed stars into twelve unequal constellations forming the background of the passage of the sun, moon, and planets, is utilized by astronomers for the purpose of locating the planets on the celestial sphere. On the other hand, the tropical zodiac, in which the cycle of the year is divided into twelve zodiacal months, originally provided a useful calendar directly related to the passage of the seasons (Figure 2).

Figure 2
Tropical Zodiac or Calendar of the Seasons

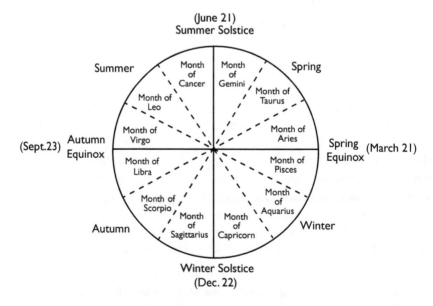

In the tropical zodiac or calendar the four seasons—Spring, Summer, Autumn, and Winter—are each divided into three equal months or signs, as shown in Figure 2.

Thus, there are two quite distinct uses of the word zodiac in contemporary culture: that of astronomers as a frame of reference for locating the planets spatially in terms of the constellations of the zodiac girding the celestial sphere, and that of the general public which continues to refer to the tropical zodiac as a division of the year applying to the passage of the seasons. However, a study of the history of astronomy reveals that there is yet a third occurrence of the term *zodiac*, which was known and employed by the ancient Egyptians, Babylonians, and Greeks. It is still used to this day by Indian astronomers. This third zodiac is a division into twelve *equal signs* more or less coinciding with the astronomical constellations of the same names. The original research presented in this thesis concerns the definition of this third zodiac. In order to distinguish it from the astronomical zodiac and the tropical zodiac, this third zodiac is referred to as the *sidereal zodiac*.[1] This nomenclature serves to distinguish it from the astronomical zodiac, which comprises twelve *unequal* constellations. The sidereal zodiac, on the other hand, comprises twelve *equal* constellations, each 30° long, and these are therefore referred to as *signs* rather than constellations.[2]

As referred to in the Introduction, the sidereal zodiac originated with the Babylonians and thus it is sometimes called the Babylonian zodiac. Knowledge of this third zodiac was generally lost in the West until the end of the nineteenth century when it was rediscovered by Joseph Epping.[3] As mentioned already in the Introduction, in the twentieth century the researches of Franz Kugler,[4] Otto Neugebauer,[5] B.L. van der Waerden,[6] Abraham Sachs,[7] Peter Huber,[8] and others, once more brought it to light, enabling it to be defined more precisely.

This thesis concerns the *intrinsic definition* of the Babylonian sidereal zodiac, i.e. how it was conceived of by the Babylonians themselves, rather than—as in

---

1. Rochberg (1999), p 45, refers to the modern tropical versus the Babylonian sidereal zodiac. *Sidereal* means 'of the stars', so the sidereal zodiac could also be termed the "zodiac of the stars".

2. Hipparchus, *Commentary to Aratus* refers to 'zodiacal signs' meaning *any* 30° arc on the celestial sphere. *Hipparchi in Arati et Eudoxi Phenomena Commentarium* (1894).

3. Epping-Strassmaier (1893), p169, n2.

4. Kugler, *Die Babylonische Mondrechnung*, pp 73-74; cf. also SSB i, pp 27-31 for the first attempt to define the Babylonian sidereal zodiac.

5. Neugebauer (1942).

6. Van der Waerden (1953).

7. Sachs (1952, 2).

8. Huber (1958).

the case of Kugler, van der Waerden, and Huber—a heuristic definition based on analysis of Babylonian observations and computations. The central thesis explored here is that the Babylonians' point of departure in defining the zodiac was the location of the bright stars Aldebaran and Antares, which are diametrically opposite one another, each in the center of their respective constellations. This striking juxtaposition led to these two stars being chosen as the fiducial axis of the Babylonian zodiac, with Aldebaran located at 15° Taurus and Antares at 15° Scorpio. Once this fiducial axis was specified, the positions of other bright stars around the zodiac could be determined, for example Regulus (α Leonis) at 5° Leo and Spica (α Virginis) at 29° Virgo.[1]

Isaac Newton, looked upon by many as the founder of modern astronomy, referred to a fixed-star zodiac comprising twelve equal (30°) signs.[2] Appendix II examines Newton's sidereal zodiac and shows that it was almost identical to that of the Babylonians, apart from a one degree difference. Newton also applied this sidereal zodiac in his chronological research. He devoted much of the latter period of his life—some thirty or forty years—to the task of showing that the zodiac with equal length (30°) constellations might be of use in the field of astronomical chronology.[3]

Appendix II investigates Newton's research underlying his claim that the sidereal zodiac, once securely defined, may have a use for historians for chronological purposes. In the main part of this work, however, research is presented which is concerned with the historical investigation of the zodiac and with the definition of the Babylonian sidereal zodiac, for which purpose a star catalogue has been compiled (Appendix I).

THE HISTORICAL APPROACH

In order to have a clear understanding of the different uses of the term *zodiac*, it is helpful to consider the historical origin of the divisions already mentioned. First let us consider in the history of astronomy the origin of the division into the unequal zodiacal constellations of the astronomical zodiac and then the origin of the tropical division of the year into twelve months commencing with the vernal equinox. Afterwards let us consider how the equal division sidereal zodiac originated. Of course, such origins may date back into a dim and distant past no longer accessible to historical research.

---

1. Appendix I provides an exact reconstruction of a star catalogue for the epoch –100 in which the longitudes of stars are indicated within the framework of the Babylonian sidereal zodiac.
2. Newton, *The Chronology of Ancient Kingdoms Amended*, p 82.
3. Manuel, *Isaac Newton Historian*, pp 16ff.

Nevertheless, there was a time when these divisions became scientifically defined, and it is in this sense that we can refer back to the originators of the various divisions. That is, the originator of a division may be thought of in the sense of the individual who first defined the division in a scientific way; and owing to the enormous advance of knowledge in the domain of the history of astronomy in the twentieth century, the individuals concerned may sometimes be identified with a high degree of certainty.

Who, then, was the originator of the division now employed in modern astronomy—the division of the fixed stars of the zodiac into twelve unequal groupings? What is the history of the division into twelve zodiacal constellations (astronomical zodiac)? And how were the boundaries between the constellations decided?

In relation to the latter question, let us take as a starting point the congress of astronomers of the International Astronomical Union (IAU) held in 1928.[1] They met to consider this very question. Until this meeting there was no universal agreement between astronomers about the exact location of the boundaries between the constellations. The outcome of the meeting was an agreement on how the boundaries are to be defined in relation to the fixed stars comprising the various stellar constellations.

> [T]he never sharply defined boundaries of the pictorial constellations were replaced (by international agreement in 1928) by a system of arcs of constant right ascension and declination (for the equinox of 1875).[2]

Thus in 1928 the boundaries of the twelve constellations of the zodiac—and also the boundaries of the other 76 extra-zodiacal constellations—were formally defined by the IAU. But what was the situation prior to 1928? What in fact was the basis for the IAU formal definition of the constellations of the zodiac?

The original basis for the IAU definition of the twelve zodiacal constellations—and also for the 36 classical non-zodiacal constellations—is provided by Claudius Ptolemy's description of the location of the principal stars in each of the 48 stellar constellations described by him in his star catalogue from AD 138. The following is a brief summary of the later transmission of Ptolemy's description of the constellations when they became drawn up as celestial atlases or star maps in the form familiar to us today:

---

1. *Transactions of the International Astronomical Union* III (1929), p13.
2. Neugebauer, HAMA III, p1087; cf. Delporte, *Délimitation scientifique des constellations* and Delporte, *Atlas Céleste* for the precise specifications of the boundaries of the constellations.

18 HISTORY OF THE ZODIAC

The situation of the principal stars in each one of the forty-eight classic constellations are verbally described by Ptolemy. In Lalande's *Bibliographie Astronomique* we find that in AD 1515 Albrecht Dürer published two star maps, one of each hemisphere, engraved in wood, in which the stars of Ptolemy were laid down by Heinfogel, a mathematician of Nuremburg. The stars themselves were connected by constellation figures drawn by Dürer. These constellation figures of Dürer, with but few changes, were copied by Bayer in his *Uranometrie* (1603); by Flamsteed in the *Atlas Coelestis* (1725); by Argelander in the *Uranometrie Nova* (1843); and by Heis in the *Atlas Coelestis Novus* (1872); and have thus become classic.[1]

PTOLEMY AND HIPPARCHUS

From the above summary it is evident that the division into constellations may be traced back to Ptolemy. Essentially the divisions of the 48 classic constellations employed by astronomers (up to 1928) were based on Ptolemy's description in Books VII and VIII of his textbook of astronomy, the *Almagest*.[2] The IAU congress of 1928 gave a precise definition of these 48 constellations (and also of the 40 modern constellations), since it is not possible to derive an exact division into constellations from Ptolemy's star catalogue but only an approximate division, owing to the fact that Ptolemy's description is verbal (see Appendix I). Thus the astronomers referred to above—Bayer, Flamsteed, Argelander, Heis, and also others—based their star catalogues of the 48 classic constellations on that of Ptolemy given in the *Almagest*. Fortunately, because of Ptolemy's excellent descriptions, there are only slight deviations in the various star catalogues, according to the differences in each astronomer's reading of Ptolemy's description. These minor discrepancies, although not completely eradicated, were overridden by agreement at the IAU meeting of 1928 through the procedure of formally defining the boundaries of the 48 classic constellations (including the twelve zodiacal constellations) and also the 40 modern constellations. From this brief historical account, since they belong to the 48 classic constellations, it is clear that the division into twelve zodiacal constellations which is presently used by astronomers as a frame of reference for the movements of the sun, moon, and planets is actually a formal definition of Ptolemy's catalogue compiled in the second century AD.

There still remains the question as to whether Ptolemy's star catalogue was itself based on an earlier one. Was it his own independent compilation or did he simply revise an already existing catalogue? It is important to answer this

1. Allen, *Star Names*, pp 28–29.
2. Ptolemy, *Almagest* VII and VIII (trsl. Toomer, pp 341–399). Cf. also Taliaferro, pp 234–258.

question in order to ascertain whether or not Ptolemy can be considered the originator of the astronomical zodiac. Fortunately Ptolemy himself informs us of the approach he adopted in compiling his catalogue:

> The descriptions which we have applied to the individual stars as parts of the constellation are not in every case the same as those of our predecessors (just as their descriptions differ from their predecessors): in many cases our descriptions are different because they seemed to be more natural and to give a better proportioned outline to the figures described. Thus, for instance, those stars which Hipparchus places 'on the shoulders of Virgo', we describe as 'on her sides', since their distance from the stars in her head appears greater than their distance from the stars in her hands....[1]

From his own investigations into the historical background of Ptolemy's catalogue, Otto Neugebauer surmises that:

> Nobody doubts that long before Hipparchus there existed descriptions of constellations in which the relative positions of the stars were described in one way or another. Hipparchus' own references to the writings of Eudoxus are proof enough. When Ptolemy says that he had redefined the association of many single stars with respect to the traditional configurations, he adds the remark that his predecessors did not act differently. But the arrangement of stars in pictures is something other than a catalogue of stars where the main problem consists in giving accurate positions for the single stars, regardless of their grouping. Here, it seems, that Ptolemy had only one predecessor, namely Hipparchus.[2]

Although little is recorded about the life of Hipparchus, who lived in the second century BC, it is known that he was born in Nicaea in Bithynia and that he devoted some time to making observations of the fixed stars.[3] He compiled his catalogue of stars probably around 129 BC but it is now lost. However, Hipparchus' sole surviving work, the *Commentary to Aratus*, contains enough information on his fixed star observations to enable a comparison with Ptolemy's catalogue to be made. In 1925 Vogt undertook such a comparison and showed that on purely astronomical grounds Ptolemy's catalogue could not be considered merely an extension and revision of the lost Hipparchian catalogue, but that it (Ptolemy's) is an independent compilation.[4] In addition, by also taking account of the historical evidence, Otto Neugebauer considers Hipparchus' catalogue to be the first star catalogue of the constellations, and concludes:

1. Ptolemy, *Almagest* VII, 4 (trsl. Toomer, p340).
2. Neugebauer, HAMA I, p287.
3. Ibid., pp274–277.
4. Vogt (1925).

Hence, to the best of our knowledge, the first catalogue of stars, though not yet based on orthogonal ecliptic coordinates, is the Hipparchian, the second one is Ptolemy's catalogue in the *Almagest*.[1]

Since Ptolemy's catalogue of stars was to a certain extent compiled independently of the lost Hipparchian one, then the catalogue presented in the *Almagest* may be regarded as the original star catalogue upon which the astronomical zodiac is based. Thus all the available evidence points to the Greek astronomer Claudius Ptolemy as the *originator* of the division of the astronomical zodiac into twelve unequal constellations.[2] This designation of *originator* applies only in the sense of Ptolemy *having provided in his catalogue the original basis* for the first precise scientific definition of the astronomical zodiac specified by the IAU, since obviously the constellations existed and were described earlier.

EUCTEMON'S SOLAR CALENDAR

The historical approach to the problem of the origin of the astronomical zodiac yields the identification that the originator of this division was Claudius Ptolemy, who lived in Alexandria in the second century AD. In order to trace the originator of the calendar division of the year into twelve solar months—the signs of the zodiac in popular terminology—it is necessary to return to an earlier phase of development of Greek astronomy. In the light of historical research, who appears as the originator of the division commencing with the vernal equinox? (This division is known as the *tropical* division because its frame of reference is provided by the equinoctial and tropical [solstitial] points; the equinoctial points are located by the sun's position in the zodiac at the vernal and autumnal equinoxes, and the tropical [solstitial] points are determined by the zodiacal location of the sun at the summer and winter solstices.)

> The first astronomer who introduced the tropical division of the zodiac seems to have been Euctemon, who lived in Athens. In the year 432 BC Meton and Euctemon observed the summer solstice (according to Ptolemy, *Almagest* III, 1). Starting with this observation, Euctemon constructed a parapegma, i.e. a calendar in which the annual risings and settings of several fixed stars were noted. He divided the solar year into 12 'months' (not lunar, but solar months) defined by the 12 signs of the zodiac. In the month of 'Cancer' the sun dwelt in the sign

---

1. Neugebauer, HAMA I, p 288.
2. Neugebauer's very comprehensive analysis of the originality of Ptolemy's star catalogue, HAMA I, pp 280–288, and HAMA III, p 1087, leads him to conclude that "all later catalogues eventually descend from the *Almagest*."

Cancer, and so on. The first day of the month 'Cancer' was the day of the summer solstice, the first day of 'Libra' was the autumnal equinox, and so on.... Euctemon's solar year began with the summer solstice. The first five 'months' had 31 days each, the last seven 30 days.[1]

Whereas the division of the astronomical zodiac into twelve unequal constellations may be traced back to Ptolemy (second century AD), the division of the year into twelve more or less equal months named after the signs of the zodiac and commencing with the vernal equinox as the start of the first month was apparently first defined by the Greek astronomer Euctemon in the fifth century BC Euctemon's solar calendar may be represented as follows (Figure 3):

Figure 3

Euctemon's Solar Calendar, with Twelve Solar Months

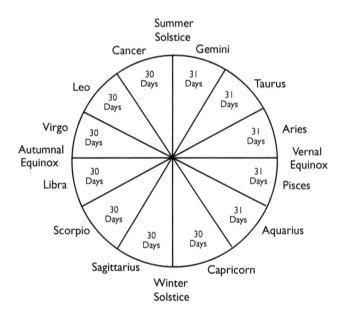

1. Van der Waerden, SA II, p289. Euctemon's parapegma was published by Rehm (1913); cf. also the translation and commentary by Pritchett-van der Waerden (1961). Although Euctemon's calendar is apparently the first to divide the year into twelve solar months taking the vernal equinox as the start of the first month, the Babylonian text MUL.APIN from the seventh century BC gives a calendar scheme of 12 months each with 30 days in which the autumnal equinox is located in the middle of the seventh month from which it follows that, in this schematic calendar, the vernal equinox should occur in the middle of the first month; cf. also Hunger-Pingree (1999), p61.

Comparing Euctemon's solar calendar with the modern tropical division of the year (see Table 1), we find agreement in the lengths of the zodiacal months Taurus, Gemini, Libra, Scorpio, Sagittarius, but that the remaining months differ by a day one way or another, and in the case of the month of Cancer, there is a two-day difference. Of course, every four years an extra day has to be inserted into the solar calendar to make up the total number of days in the year from 365 to 366, in order to ensure that the first day of the month of Aries coincides with the day of the vernal equinox, and that the first day of the month of Cancer coincides with the day of the summer solstice, and so on. In the civil calendar it is customary to add in this day at the end of the month of February, while in the solar calendar the lengths of the months can vary by one day, more or less, from year to year, according to the dates on which the equinoxes and solstices fall. However, since no historical application of Euctemon's calendar has been found, it is not known if there was any rule for the addition of an extra day every fourth year. The lengths of the months given in Table 1 can be taken as a guide to a schematic representation of a modern solar calendar (in the same manner as Euctemon's calendar is represented in Figure 3), and we arrive at the following scheme for a modern, more exact specification of a solar calendar (Figure 4).

In Figure 4 the five months preceding the autumnal equinox in the modern solar calendar are *long*, while the seven solar months following the autumnal equinox are *short*. Hence the schematic form of Euctemon's calendar—five months of 31 days, followed by seven months of 30 days—is comparable with the scheme of the modern solar calendar. The difference between them is simply that the central month of the five long ones, Aries in Euctemon's scheme, has now become Cancer in the modern scheme.

The reason the month of Cancer is at the center of the long months is that during the month of Cancer the earth is furthest from the sun (earth at aphelion), and hence is moving more slowly along its elliptical orbit around the sun. Correspondingly, shortly after Christmas in the month of Capricorn, the earth approaches nearest to the sun (earth at perihelion), and is then moving most rapidly along this part of its elliptical orbit. The date (around January 3) on which the earth is at perihelion equates with (approximately) the thirteenth day of the solar month of Capricorn, while the date (around July 4/5) on which the earth is at aphelion equates with (approximately) the fourteenth day of the solar month of Cancer. In other words, the line of the apsides (aphelion-perihelion) is crossed by the earth as it orbits the sun at approximately the middle of the solar months Cancer and Capricorn during the present epoch of time.

SIGNS AND CONSTELLATIONS OF THE ZODIAC

Figure 4

Schematic Representation of a Modern Solar Calendar

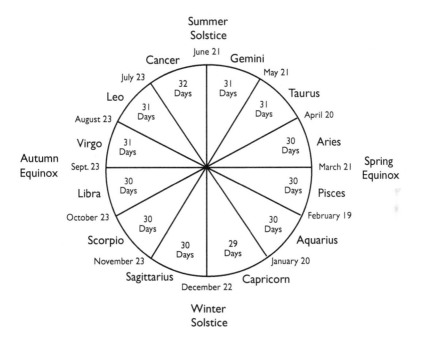

In Euctemon's time the situation was otherwise. Then, if Euctemon's scheme were to have been astronomically accurate, the earth ought to have crossed the apsidal line somewhere around the middle of the solar months of Aries and Libra. However, closer investigation shows that the calendar of Euctemon did not fit the facts accurately. Neugebauer remarks that Euctemon's scheme "has nothing to do with the solar anomaly, as is evident, e.g. from the direction of the corresponding 'apsidal line' through Aries and Libra...."[1] Indeed, computation shows that the apsidal line in the fifth century BC—around the time of Euctemon (432 BC)—lay in the direction through Gemini and Sagittarius, with the earth being at the furthest distance from the sun on (approximately) the second day of the solar month of Gemini. The long months should therefore have been centered about Taurus/Gemini and not Aries, for Euctemon's scheme to fit the astronomical facts.

1. Neugebauer, HAMA II, p 628.

It is different with the calendrical scheme of Callippus, as this has been reconstructed by Rehm.[1] Callippus, a contemporary of Aristotle (fourth century BC), devised an accurate parapegma, with Gemini placed at the center of five long months and Sagittarius occupying the central position of seven short months (see Table 3).[2] In his scheme, by symmetry, the apsidal line lies in the direction through Gemini and Sagittarius, which was actually the case in the fourth century BC. When Aristotle died (322 BC) Callippus was about fifty years old, and the earth reached its furthest distance from the sun on about the fourth day of the solar month of Gemini.

After that of Callippus, the next and possibly best known astronomical scheme of the seasons utilizing solar months which has been fully reconstructed, is the so-called Geminus parapegma. This is appended to Geminus' *Introduction to the (celestial) Phenomena*, an elementary astronomical treatise known as the *Isagoge*. Although the *Isagoge* was written about AD 50, the calendric text following it is probably two centuries or so older.[3] Unlike both Euctemon's and Callippus' calendars, that of Geminus is not symmetric, and hence does not allow us to determine the location of the apsidal line as the axis of symmetry. From Euctemon's scheme it can be deduced that the apsidal line passed through Aries and Libra, which was grossly inaccurate for the time at which he lived. However, Callippus devised a scheme that correctly placed the line of the apsides through Gemini and Sagittarius, which was accurate in his time (fourth century BC). The Geminus parapegma (around the second century BC) should also have had the same format as that of Callippus in order to be correct, but from the asymmetric distribution of the 365 days of the year among the twelve solar months it can be seen that there is a discrepancy (Table 3).

Table 3

Comparison of the Solar Calendars of Callippus and Geminus

	Arie	Taur	Gemi	Canc	Leo	Virg	Libra	Scor	Sagit	Capr	Aqua	Pisc
Callipus	31	31	32	31	31	30	30	30	29	30	30	30
Geminus	31	32	32	31	31	30	30	30	29	29	30	30

1. Rehm, RE Par., col. 1346f.
2. Pritchett-van der Waerden (1961), pp39–41, concerning parapegmata; cf. also Rehm (1913). Note that a parapegma, as referred to already, is a calendar in which the annual risings and settings of several fixed stars were noted.
3. Neugebauer, HAMA II, pp579–581 and p587.

From the foregoing it is evident that in Greek astronomy there was a tradition of *parapegmata*, into which were incorporated solar calendars.¹ This tradition in Greek astronomy appears to have originated with the astronomical school of Meton and Euctemon (fifth century BC) based in Athens. As van der Waerden concludes: "The first astronomer who introduced the tropical division of the zodiac seems to have been Euctemon...."² It is not unreasonable, therefore, to regard Euctemon as the originator of the division of the year into twelve solar months named after the signs of the zodiac. For the sake of clarity, however, the division of the year into twelve solar months originating with Euctemon in the fifth century BC should be called the *tropical calendar*. In fact, as discussed in Chapter 3, the *tropical zodiac* (as distinct from the tropical calendar) evidently originated much later, probably with Hipparchus, in the second century BC. Even if Euctemon (and also his colleague Meton) knew of zodiacal signs, as the following statement indicates, it can only be the signs of the Babylonian sidereal zodiac about which he (they) knew, since the innovation of the ecliptic coordinate system of the tropical zodiac, associated with Hipparchus, took place some three centuries later.

> It is practically certain that Euctemon used zodiacal signs. Now, the names of most of the signs, as we know them from Greek texts, are literal translations of Babylonian names of zodiacal constellations. The Babylonians used the same names to designate zodiacal signs of exactly 30 degrees, just as the Greeks did. Cleostratus of Tenedus, who probably lived about 500 BC, is said to have introduced the zodiacal signs into Greece. In any case, the names of the signs (like Taurus, Gemini, Leo, Virgo, Libra, Scorpio, Sagittarius, Capricorn and Pisces) clearly show that the signs are not an independent Greek invention.³

From the preceding discussion it can be reasonably concluded that Euctemon devised a solar calendar (the tropical calendar) based on the division of the year occasioned by the solar phenomena of the solstices and the equinoxes, and that he named the twelve solar months of this calendar after the twelve signs or constellations of the zodiac. According to John Britton, the twelve signs of the Babylonian sidereal zodiac "appeared between −463 and −453."⁴

---

1. Cf. Pritchett-van der Waerden (1961), pp 45–47, for a discussion of the relation between Greek parapegmata and the calendrical scheme known from cuneiform sources in the compendium MUL.APIN, which was composed at Babylon in the seventh century BC.
2. Van der Waerden, SA II, p 281. It would be more accurate to say that, "The first astronomer who introduced the *tropical calendar* seems to have been Euctemon." For the history of the *tropical zodiac*, which did not originate with Euctemon but was probably introduced into astronomy by Hipparchus, see chap. 3.
3. Pritchett-van der Waerden (1961), p 45.
4. Britton (1999), p 244.

Theoretically, therefore, it is conceivable that twenty or thirty years after the origin in Mesopotamia of the equal division signs of the zodiac (in contrast to the unequal constellations of the zodiac) Euctemon in −431 could have known about the zodiacal signs, assuming that there was a certain amount of cross-cultural transmission between Greece and Babylonia in the fifth century BC.[1] And it is reasonable to suppose that Euctemon named the months of the tropical calendar after the twelve constellations of the zodiac according to the principle of correspondences between 'above and below', which was characteristic of the Greek tendency in antiquity to reason by way of analogy. In other words, the twelvefold division of the time cycle of the year given by the tropical calendar was probably seen by Euctemon to correspond by way of analogy with the twelvefold division of the celestial sphere visible in the twelve zodiacal constellations. And since the great circle of the twelve constellations on the celestial sphere was seen to begin with Aries, where the sun was located on the day of the vernal equinox, the first solar month of nature's yearly cycle indicated by the tropical calendar was named after Aries. Further, it may be also conjectured that the subsequent solar months were named after the following constellations of the zodiac accordingly. Hence it seems likely that the tropical calendar as a twelvefold division of time arose by way of analogy with the twelvefold division of the celestial sphere visible in the circle of the twelve zodiacal constellations. The same line of reasoning applies, of course, if it was the equal division *signs* of the Babylonian sidereal zodiac to which this analogy was applied rather than the unequal zodiacal constellations. It is interesting to note that the astronomical school of Meton and Euctemon evidently used two systems simultaneously: Euctemon's tropical calendar defining the solar year and the Babylonian sidereal zodiac with the vernal point at 8° Aries (System B) as indicated by the juxtaposition of the two following statements:

> Euctemon constructed a parapegma, i.e. a calendar in which the annual risings and settings of several fixed stars were noted. He divided the solar year into 12 'months' (not lunar, but solar months) defined by the 12 signs of the zodiac. In the month of 'Cancer' the sun dwelt in the sign Cancer, and so on. The first day of the month 'Cancer' was the day of the summer solstice, the first day of 'Libra' was the autumnal equinox, and so on....[2]

---

1. According to Porphyry in his *Life of Pythagoras*, Pythagoras was the first to introduce the teachings of the Chaldeans into Greece, after having gone to Babylon where he associated with the other Chaldeans, especially attaching himself to Zaratas [=Zoroaster].... It was during his stay among these foreigners that Pythagoras acquired the greater part of his wisdom" (trsl. K.S. Guthrie, p125).
2. Van der Waerden, SA II, p289.

These two astronomers observed the summer solstice in Athens in 432 BC (Ptolemy, *Almagest* III, 1).... Meton placed the equinoxes and solstices at 8 degrees Aries, 8 degrees Cancer, etc. (Columella, *De re rustica* IX, 14), exactly like the Babylonian System B calculations.[1]

The midsummer sun for Meton and Euctemon (432 BC) was in the constellation of Cancer (Meton specified it to be at 8° Cancer whereas—as indicated in Appendix I—it was actually at 9° Cancer in the Babylonian sidereal zodiac at that time). At the present time, on account of precession, the midsummer sun is located in the constellation of Gemini (currently at 5° Gemini in the Babylonian sidereal zodiac). From this it is readily apparent that the two astronomical systems—that of the Babylonians (the sidereal zodiac) and the tropical calendar introduced by Euctemon—are intrinsically independent and complement one another.

THE PROGRESSION OF THE APSIDES

From the above analysis concerning the origin of the division of the year into twelve solar months, it emerges that the apsidal line is engaged in a slow progressive motion in relation to the solar months of the tropical calendar. In the fourth century BC Callippus devised a calendrical scheme in which the apsidal line lay in the direction of Gemini and Sagittarius, which was correct for his epoch. By the middle of the thirteenth century AD, however, the scheme of Callippus was no longer valid. Computation shows that in approximately AD 1250 the apsidal line coincided with the tropical points (summer solstice and winter solstice). That is, in the middle of the thirteenth century the day of the year on which the earth lay furthest from the sun (earth at aphelion) coincided with the longest day of the year, the day of the summer solstice.[2] From the middle of the thirteenth century onwards the apsidal line was located in the direction of Cancer and Capricorn. The progression has been such that in our time the day on which the earth crosses this line as it orbits the sun, and when the distance between the sun and earth is at a maximum, is near the middle of the zodiacal month of Cancer—about thirteen days after the summer solstice.

This slow motion of the apsidal line was unknown to the ancients.[3] It was first discovered by the Islamic astronomer, Thābit ben Qurra, in the ninth

---

1. Ibid., p246.
2. Throughout the above discussion of Euctemon's solar calendar it is only the northern hemisphere that is being considered.
3. Neugebauer, HAMA I, p421, n4.

century. Until his discovery it was believed that the sun arrived at apogee[1] on the sixth day of the month of Gemini, as had been determined by Ptolemy in the second century AD.[2] Thus one reason why the motion of the apsidal line was not discovered until the ninth century had to do with Ptolemy's authority, which generally went unquestioned. Another reason why the apsidal motion was not discovered until later is that it could only be detected indirectly through comparison of the small differences in lengths of the months of the tropical calendar.

Now it is known that the progression of the apsidal line completes a revolution in relation to the tropical points in about 21,000 years.[3] Thus the astronomical situation prevailing in the middle of the thirteenth century, when the earth reached aphelion on the first day of the month of Cancer (the day of the summer solstice), will be repeated in ca. AD 22,250, and previously occurred in ca. 19,750 BC Moreover, approximately midway through the 21,000-year interval extending from 19,750 BC to AD 1250, i.e. in ca. 9250 BC, the revolution of the apsidal line in relation to the tropical points gave rise to exactly the opposite situation to that of the mid-thirteenth century. In ca. 9250 BC the day on which the earth arrived at aphelion, at greatest distance from the sun, coincided with the winter solstice, and this situation will be repeated in ca. AD 11,750 (Figure 5).

Since the period of the progression of the apsides in relation to the twelve zodiacal months is about 21,000 years, the corresponding length of time for its passage in relation to one solar month is approximately 1750 years,[4] e.g. from 500 BC to AD 1250 for the solar month of Gemini, from AD 1250 to 3000 for the solar month of Cancer, etc., are the (approximate) historical periods during which the earth is at maximum distance from the sun during the given month of the tropical calendar.

To summarize: although the progression of the apsides is not directly relevant to the *definition* of the zodiac, it is evident that the discovery of the motion of the apsidal line was only able to be made on the basis of a consideration of the lengths of the solar months of the tropical year. Thus the introduction into Greek astronomy of the solar months by Euctemon in the fifth century BC was a prerequisite for this discovery. As we shall see, already prior

---

1. In the geocentric astronomical system the sun is said to be at apogee when the distance between the earth and the sun is at a maximum, i.e. when the earth is at aphelion heliocentrically.

2. Neugebauer, HAMA I, pp 55–61. Here Ptolemy's tropical longitude for the apogee ♊ 5°30', i.e. 95°30' from the vernal point, is equated with the sixth day of the month of Gemini.

3. This period of revolution is difficult to estimate with any degree of precision because the rate of motion of the apsidal line is varying; it appears to follow a sinusoidal curve.

4. 1750 years is one-twelfth of 21,000 years.

to Euctemon the Babylonians had arrived at a definition of solar months, and in the following we shall return to the question as to the relationship between Euctemon's calendar and its fore-runner defined by the Babylonians more than 2½ centuries earlier.

Figure 5

The Progression of the Apsides
(Approximate Dates)

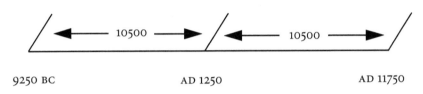

9250 BC   AD 1250   AD 11750

9250 BC—*earth at aphelion coincided with the winter solstice*
AD 1250—*earth at perihelion coincided with the winter solstice*
AD 11750—*earth at aphelion will coincide with the winter solstice*

2

THE PRECESSION OF THE EQUINOXES

IN CONSIDERING THE MOVEMENT of the apsidal line in relation to the tropical points we are concerned solely with the changing relationship between the sun and the earth. In this movement the change becomes noticeable over long periods of time. At different epochs of time the distance between the sun and the earth alters for the same (given) time of the solar year which, as described in Chapter 1, may be divided into twelve solar months corresponding to the twelve signs or constellations of the zodiac. At the present time the earth is furthest from the sun during the solar month of Cancer,[1] but from the fifth century BC until the middle of the thirteenth century AD it was during the solar month of Gemini that the earth was at its maximum distance from the sun (earth at aphelion). This gradually changing relationship—the progression of the apsidal line in relation to the zodiacal months—has a period of approximately 21,000 years (Figure 5).

Like the motion of the apsidal line, the precession of the equinoxes is a slowly changing astronomical variable, specified by the retrograde movement of the vernal point through the constellations of the zodiac. It has a period of about 26,000 years.[2] This is of the same order as the period required for the revolution of the apsidal line in relation to the solar months. Although the precession of the equinoxes is a slow-moving phenomenon, as with the progression of the apsides it becomes gradually perceptible over the course of centuries. Whereas the motion of the apsidal line was not discovered until the

---

1. Cf. figures 3 and 4 defining the solar months. Thus the solar month of Cancer is the fourth solar month of the year, which (in the calendar of twelve solar months) begins with the vernal equinox. The solar month of Cancer begins with the day of the summer solstice (= Day 1).

2. Using Newton's constant of precession, 1° in 72 years, the equinoxes retrogress through the 360° of the zodiac in 360 x 72 = 25,920 years.

ninth century AD,[1] the precession of the equinoxes was discovered a thousand years earlier by Hipparchus in the second century BC.[2] As mentioned briefly in Chapter 1, one reason astronomy had to wait so long for the discovery of the progression of the apsides is because—unlike precession, which was observed directly—the motion of the line of the apsides could only be observed indirectly from the small differences in the lengths of the solar (tropical) months. For Ptolemy the apsidal line was fixed with respect to the solstices with the solar apogee at 5♊30,[3] corresponding to the sixth day of the month of Gemini. Eventually, however, astronomers could not fail to notice that Ptolemy's placing of the solar apogee was too far away from the summer solstice to fit the facts. It is to the credit of Thābit ben Qurra that he was the first astronomer to observe this discrepancy.

Hipparchus was able to make his discovery of precession because not only was he a careful observer of the stars, but also he had at his disposal some observations made almost two centuries earlier—between 295 BC and 283 BC by Timocharis, another Greek astronomer.[4] By comparing his own observations with those of Timocharis he discovered that there had been a movement of the autumnal point 2° westward (Figure 8) in relation to the star Spica (α Virginis).

> For Hipparchus too, in his work *On the displacement of the solstitial and equinoctial points*, adducing lunar eclipses from among those accurately observed by himself, and from those observed earlier by Timocharis, computes that the distance by which Spica is in advance of the autumnal [equinoctial] point is about 6° in his own time, but was about 8° in Timocharis' time.[5]

This observation led Hipparchus to conclude that the tropical and equinoctial points *precess* through the zodiac at a rate of at least 1° in a century.[6] The discovery of this westward shift of the equinoctial points through the zodiac is unquestionably "Hipparchus' most famous achievement."[7]

Ptolemy utilized the observations of Timocharis and Hipparchus. These together with his own observations enabled him to confirm Hipparchus' discovery.

---

1. By Thābit ben Qurra (died 901); Neugebauer, HAMA I, p 58.
2. Neugebauer, HAMA I, pp 292–296.
3. Neugebauer, HAMA I, p 307. In addition, from the point of view of practical astronomy, the apsidal motion is rather more difficult to detect than the precession of the equinoxes, since greater observational accuracy is required.
4. Ptolemy, *Almagest* VII, 3 (trsl. Toomer, pp 329–338).
5. Ibid., VII, 2 (trsl. Toomer, p 327).
6. Ibid. (trsl. Toomer, p 328).
7. Neugebauer, HAMA I, p 292.

32 HISTORY OF THE ZODIAC

From this we find that 1° rearward [eastward] motion [of the fixed stars in relation to the tropical and equinoctial points] takes place in approximately 100 years.[1]

Ptolemy's rate of precession of 1° in 100 years was accepted until the ninth century when the independent observations of Muslim astronomers showed that this was too slow.[2]

THE PHENOMENON OF PRECESSION

Equinoctial precession is a phenomenon that is independent of the movement of the apsidal line. The latter movement has been considered in Chapter 1 in relation to the twelve solar months of the *tropical calendar*. The precession of the equinoxes, however, must be considered in relation to the *fixed stars*. Moreover, the factor of distance between the sun and the earth—the factor determining the location of the apsides—can be ignored. The element which plays the all-important role in the determination of precession is the location of the sun against the background of the fixed stars of the zodiac on the day of the vernal equinox. The actual moment of the equinox is defined by the sun's crossing the celestial equator, moving northward, at the beginning of spring (Figure 6).

At the present time the location of the vernal point is to be found just below the stars marking the tail of the southwestern Fish of Pisces (Figures 1 and 6). This identifies the location of the sun on the celestial sphere at the start of the solar year. Owing to precession this location alters slowly with time. For example, according to Columella, at the time of Meton and Euctemon in the fifth century BC, the vernal point was in the constellation of the Ram (East of the Fishes) at 8° Aries-Ram.[3]

By computing backwards from the present time to Meton's time using Newton's constant of precession of 1° in 72 years, the vernal point in 432 BC is found to be directly beneath the star Sheratan (β Arietis) marking the western horn of the Ram. Thus in relation to the commencement of the solar year, i.e. the vernal equinox or the first day of the solar month of Aries in the tropical calendar, the sun now occupies a quite different region of space than it did in 432 BC.

---

1. Ptolemy, *Almagest* VII, 2 (trsl. Toomer, p 328).
2. Neugebauer, HAMA I, p 293: al-Battānī (858–929) concluded that the rate of precession was 1° in 66 years.
3. Columella, *De re rustica* IX, 14 (trsl. Ash-Forster–Heffner, p 487).

Figure 6

The Present Location of the Vernal Point

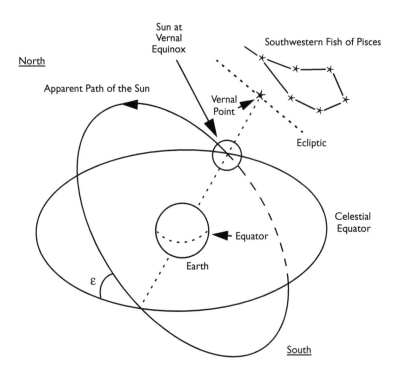

Definitions
Celestial equator = projection of earth's equator.
Ecliptic = projection of the apparent path of the sun.
ε, the obliquity of the ecliptic = angle between celestial equator and ecliptic (ε = 23½°).
Vernal point = projection onto ecliptic of extended earth–sun line at moment of vernal equinox. At present the vernal point is located in the southwestern Fish of Pisces.

The vernal point has shifted westward through the zodiac by approximately 33¾° between 432 BC and the present time.[1] It no longer occupies the sector of space known as Aries-Ram but is to be found now in the region known as Pisces-Fishes at the tail of the southwestern Fish, a little to the west of and beneath the star ω Piscium (Figure 7).

---

1. 432 BC to AD 2000 is 2431 years; and using a rate of precession of 1° in 72 years, then in 2431 years this amounts to approximately 33¾°, since 33¾ × 72 = 2430.

Figure 7

Precession of the Spring Equinox Between 432 BC and AD 2000

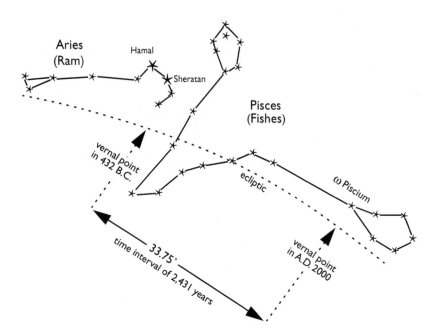

Owing to the changing location of the sun's position in space at the start of the solar year, the vernal point has retrogressed along the ecliptic through 33¾° since 432 BC. This motion is known as the precession of the equinoxes. In theory it is identified by observing the sun's position in space on the day of the vernal equinox at different epochs of time. Since the position of the sun against the background of the fixed stars can only be observed directly when there is a solar eclipse,[1] the direct observation of precession might theoretically be possible from a comparison of solar eclipse records on a given day of the solar year, e.g. at the spring equinox, on day one of the solar month of Aries. In practice, however, there are too many problems for this possibility to be realized. Alternatively, and this was the method of Hipparchus, since lunar eclipses occur when the moon is on or near the ecliptic in a position diametrically opposite to that of the sun on the celestial sphere, then precession can also be determined from a comparison of lunar eclipse records on a given day

---

1. This is the origin of the term 'ecliptic'. It is the locus of points through the zodiacal constellations at which eclipses can occur.

of the solar year. Hipparchus discovered precession by comparing an observation that he had made of the moon's position in relation to Spica at a lunar eclipse at the time of the vernal equinox with a similar observation made by Timocharis a little less than two centuries earlier.[1]

THE RATE OF PRECESSION: HISTORICAL SURVEY

From the foregoing considerations it is evident that if attention is given not only to the passage of the solar months in the tropical calendar, but also to the sun's position on the celestial sphere at the commencement of the solar year, then the gradual shift of the sun's position back through the sidereal zodiac is discovered. This movement is called the *precession* of the equinoxes since it is the motion of the vernal point, the sun's ecliptic location on the day of the spring equinox, *backwards* through the fixed-star zodiac. It can be determined astronomically that since the time of the Greek astronomers Meton and Euctemon, over 2430 years ago, precession has amounted to about 33¾°. This amount assumes a constant rate of precession of 1° in 72 years which, rounded to the nearest year, is the modern accepted mean value that was determined in the seventeenth century by Newton.[2]

Prior to Newton, Copernicus stated that it was 1° in 71 years during the centuries preceding his own time, but he believed the rate of precession to be variable, with different values at earlier epochs of time.

> Hence in those 400 years before Ptolemy, clearly the precession of the equinoxes was slower than from Ptolemy to al-Battānī, when it was also quicker than from al-Battānī to our times.[3]

Copernicus knew that al-Battānī's parameter for precession was 1° in 66 years, and that Ptolemy had determined the rate of precession to be 1° in 100 years.[4] Instead of concluding that either of these astronomers might have been in error, Copernicus accepted their precessional coefficients at face value. He deduced that prior to Ptolemy (second century AD) precession was 1° per century, while after Ptolemy and before al-Battānī (858–929) it was 1° in 65 years.[5] Thus in comparison with his own value of 1° in 71 years, al-Battānī's rate of precession is fast and Ptolemy's is slow. Since Ptolemy's estimate of 1° per century was based on observations made over a 400-year period prior to the second

---

1. Ptolemy, *Almagest* VII, 2 (trsl. Toomer, p327).
2. Cajori, *Newton's Mathematical Principles*, p490.
3. Copernicus, *De revolutionibus* III, 2 (trsl. Rosen, p122).
4. Ibid.
5. Ibid.

century AD and al-Battānī's value of 1° in 66 years took account of observations made up to the ninth and tenth centuries whilst Copernicus' own parameter, 1° in 71 years, included observations made up to his time (sixteenth century), Copernicus concluded that the rate of precession had been slow in the period before Ptolemy, that it had speeded up in the time between Ptolemy and al-Battānī, and that it had settled down to a steady rate in the time between al-Battānī and himself.[1]

Ptolemy's precessional constant of 1° in 100 years was generally accepted by medieval astronomers. It was also the limiting value determined by Hipparchus.[2]

> In the work *On the length of the year* Hipparchus came to the conclusion that the equinoctial points move at least 1° per century in a direction opposite to the order of the zodiacal signs. This is the famous constant of precession which was considered valid until the ninth century when Muslim astronomers began to make independent observations.[3]

This historical survey of astronomical knowledge concerning the rate of precession indicates that from the time of the discovery of precession by Hipparchus until the time of the Arabic astronomers, the estimated value was too small ('slow'), and this (Ptolemy's) value of 1° per century was generally accepted. The Arabic astronomers over-compensated for Ptolemy's value and arrived at a 'fast' constant of precession. Eventually, Copernicus arrived at a value of 1° in 71 years and then Newton, with more accurate observations and a more sophisticated mathematical astronomy, arrived at a value of 1° in 72 years.[4] This survey of precession is important for an understanding of the Babylonian sidereal zodiac, as will become evident, since the sidereal zodiac, once it is accurately specified, offers a background frame of reference for the movement of the vernal point as it retrogresses through the zodiac (Appendices I and II).

---

1. Ibid.
2. Ptolemy, *Almagest* VII, 2 (trsl. Toomer, p328).
3. Neugebauer, HAMA I, p293.
4. Modern astronomy has determined that the rate of precession varies slightly during the course of a complete cycle of precession lasting approximately 26,000 years. The conclusion of modern astronomy is that the average rate of precession over a complete cycle is close to 1° in 72 years, but at the present time it is slightly faster than this value. At the present time it is 1° in 71.6333 years, a value between that of Copernicus and Newton.

HIPPARCHUS' DISCOVERY OF THE PRECESSION OF THE EQUINOXES

From the *Almagest* it is evident that Hipparchus discovered precession in the latter part of the second century BC.[1] He compared an observation of a lunar eclipse made by Timocharis with his own observation of a lunar eclipse made at the same time of the solar year, the vernal equinox. Since Timocharis' observations were made between −294 and −283, it is possible to compute the lunar eclipses that occurred during those years and to assess with a high degree of certainty which eclipse observation it was that Hipparchus utilized. By computation it is found that a lunar eclipse took place during the night 17/18th March, −283, with the eclipsed moon appearing just east of Spica. The 18th March, −283, equates with the 24th day of the solar month of Pisces, i.e. six days before the vernal equinox.[2] In all probability it was this lunar eclipse which Timocharis observed and which Hipparchus subsequently made use of.

Similarly if the lunar eclipses occurring during the period that Hipparchus was making observations are examined, the most likely lunar eclipse upon which Hipparchus based his discovery of precession is that which occurred during the early hours of 21st March in −134. Again the eclipsed moon was just east of Spica; and 21st March in −134, equates with the 27/28th day of the solar month of Pisces, just two or three days prior to the vernal equinox. Moreover, "all accessible data for Hipparchian fixed star observations suggest the decades from −150 to −130,"[3] into which period the lunar eclipse at the vernal equinox of −134 falls. This also coincides with the so-called 'Hipparchus' star', which Pliny asserts was the motivation for Hipparchus to compile his catalogue of stars.[4] According to Matouan-lin, a Chinese historian of the thirteenth century, there was an 'extraordinary star' which appeared in the region of the 'claws' of the Scorpion, and which is dated to July −133.[5] The astronomer John Herschel thought that 'Hipparchus' star' was possibly a Nova.[6] In this case the

---

1. Ptolemy, *Almagest* VII, 2 (trsl. Toomer, pp 327–328).
2. The chronological utility of the days of the solar months of the tropical calendar is apparent here, since dates in any calendar can be equated with a corresponding day in a solar month, where the latter is securely defined with respect to the astronomical determination of solstices and equinoxes. It should be noted that the tropical calendar is a schematic calendar and therefore the days of the solar months in general do not coincide with actual days extending from midnight to midnight, e.g. day 1 of Aries begins at the moment of the vernal equinox and lasts until the sun has traveled exactly 1°, approximately 24 hours later.
3. Neugebauer, HAMA I, p 276.
4. Pliny, NH II, 95 (trsl. Rackham I, p 239).
5. Biot (1843), p 61; cf. also Neugebauer, HAMA I, p 284.
6. Herschel, *Outlines of Astronomy* (1849), p 563.

'extraordinary star' of July −133 might have been a Nova, which could be the phenomenon to which Pliny refers.

If the above-mentioned lunar eclipses are the ones utilized by Hipparchus in his discovery of precession, then it is evident that he noticed about 2° of precession in the period from Timocharis' eclipse in −283 and his own lunar eclipse at the vernal equinox in −134, i.e. 2° in 149 years[1] (Figure 8):

Figure 8

Retrogression of the Autumnal Point

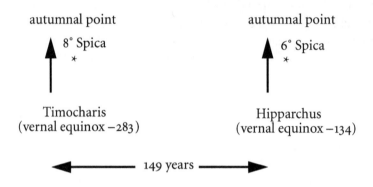

From the eclipse observations of Timocharis and Hipparchus it is evident that in the course of 149 years the autumnal point (opposite the location in the ecliptic of the sun on the day of the vernal equinox) had retrogressed about 2° in the zodiac, approaching Spica. In conformity with this observation Hipparchus concluded that the tropical (solstitial) and equinoctial points make a slow movement in a retrograde direction along the ecliptic. In fact, based on the above observations (Figure 8), Hipparchus should have concluded that the rate of precession is about 1° in 75 years, which is quite close to the modern value computed by Copernicus and Newton. From Ptolemy's citation in the *Almagest* it may be inferred that Hipparchus gave a range of values for the rate of precession, with a limiting ('slow') value of 1° in 100 years, the value later adopted by Ptolemy.[2]

As discussed below, the phenomenon of precession is important with regard to understanding the definition of the sidereal zodiac, just as the movement of the earth's apsidal line is significant in relation to the twelve solar

---

1. Kudlek-Mickler, *Solar and Lunar Eclipses of the Ancient Near East from 3000 BC to 0*, pp 152–153, indicates details of Timocharis' eclipse in −283 and Hipparchus' eclipse in −134.
2. Ptolemy, *Almagest* VII, 2 (trsl. Toomer, p 328).

months of the tropical calendar. The significance of the tropical calendar as the forerunner of the tropical zodiac is also discussed below.

THE ALLEGED BABYLONIAN DISCOVERY OF PRECESSION

From modern knowledge of Babylonian astronomy gained since the latter part of the nineteenth century through the excavation and decipherment of cuneiform texts, it might appear at first sight that the Babylonians were aware of precession. The coordinate system which they used was sidereal, in which the zodiacal belt of fixed stars was divided into twelve equal 30° sectors or signs more or less coinciding with Ptolemy's specification in the *Almagest* of the twelve constellations of the astronomical zodiac.[1] The names of many of the Greek zodiacal constellations, moreover, are more or less direct translations of the Babylonian names for the zodiacal signs.[2] From cuneiform sources it is known that the Babylonians utilized the division into twelve zodiacal signs, with each sign defined in relation to certain prominent fixed stars,[3] as early as the Achaemenid period.[4]

> That this system is of Babylonian origin became clear as soon as one gained access to the ephemerides in cuneiform texts of the Seleucid period. Here it will suffice to remark that the zodiacal signs of fixed equal length (30° each) are already attested in the Achaemenid period.[5]

It is this original zodiac of the Babylonians—the sidereal zodiac—which is of primary interest in this thesis, because it was the forerunner of all other definitions of the zodiac. In Appendix I there is an exact specification of the Babylonian zodiac in terms of a star catalogue for the epoch −100 in the framework of the sidereal zodiac. In the following chapters also the transmission of this zodiac to other cultures is traced, as far as this is possible.

Van der Waerden in his 'History of the Zodiac' (1953) claimed that the earliest reference to zodiacal signs (equal 30° divisions of the sidereal zodiac) rather than zodiacal constellations (unequal divisions of the astronomical zodiac) was in the cuneiform text VAT 4924 from the year −418.[6] However, a more recent publication from the year 1969, analyzing the tablets BM 36559

---

1. Kugler, *Die Babylonische Mondrechnung*, pp 73–74 and SSB I, pp 27–31.
2. Van der Waerden, SA II, pp 287–288.
3. Known as 'Normal-Stars'; cf. Graßhoff (1999) and Hunger-Pingree (1999), pp 148–151.
4. In a list of solar eclipses for the years from −474 to −456; cf. Aaboe-Sachs (1969), p 17.
5. Neugebauer, HAMA II, p 593. The Achaemenid dynasty ruled Babylon from −537 to −329.
6. Van der Waerden (1953), pp 220–224. This holds for the constellations of the Bull (*Bull of Heaven*), Twins (*Great Twins*), Crab, Lion, Scales, Scorpion, and Goat (*Goat-Fish*), whereas the other zodiacal constellations had different names. Thus the Babylonian equivalent of the Ram is

+ 36941 and BM 36737 + 47912, brought to the light of day the earliest recorded use of the sidereal zodiac found thus far: a cuneiform text which presents calculated and observed data for 38 possibilities of solar eclipses in the period from 5th December 475 to 21st July 457 BC.[1] The longitudes of the eclipse possibilities are given in degrees and minutes within zodiacal signs. Clearly this text was compiled after 457 BC "because it presents calculated and observed data for the 38 eclipse possibilities."[2]

Britton suggests, on the basis of references to other astronomical observations from –463 and –453, both apparently involving zodiacal signs, that the sidereal zodiac of twelve equal (30°) signs had been introduced between –463 and –453.[3] Britton's conclusion is a reasonable one, although Pingree cautions that the solar eclipse tablets might have been compiled much later in the fifth century BC and that the "alleged reference to a zodiacal sign for –463 is in fact to a constellation, while that alleged to occur ... for –453 is inconclusive."[4] Whether or not the sidereal zodiac emerged prior to –450 or somewhat later, it was definitely in use by the end of the fifth century BC, as attested to not only by the observational text VAT 4924, an astronomical diary from the year –418, which gives the positions of the planets Jupiter, Venus, Mars, Saturn, and Mercury with respect to zodiacal signs but also by the earliest Babylonian horoscopes found so far, dated 12/13th January –409 and 29th April –409.[5]

It is clear from the texts BM 36559 + 36941 and BM 36737 + 47912 relating to the period –474 to –456 and also from the text VAT 4924 from –418 that they mark a transition from the observational phase to the mathematical phase of Babylonian astronomy. This transition will be examined in detail in Chapter 5. Further, the arising of astrology as the practice of casting horoscopes for a child's birth, attested to from –409 onwards, coincided with this transition to a more sophisticated mathematical level of Babylonian astronomy, which reached a highpoint in the Seleucid era (–310 to –127).

How and when the Greeks assimilated the Babylonian sidereal zodiac is not

---

the *Hired Man* (*Hireling*), and that of the Virgin is the *Furrow*, the constellation of Sagittarius is called *Pabilsag* (corresponding to the Archer), the Water-Bearer is known as the *Great One*, and the constellation of the Fishes is referred to as *Anunītu*, cf. Hunger-Pingree (1999), p71.

1. Aaboe-Sachs (1969), p17; cf. also Hunger-Pingree (1999), pp185–200 and p222.
2. Hunger-Pingree (1999), p185.
3. Britton (1999), p244; cf. also Britton (1993). Moreover, cf. above, p6, n5, for a clarification concerning the use of the minus (–) notation for BC dates.
4. Hunger-Pingree (1999), p185.
5. Rochberg (1998), pp51–57. Pingree agrees that the planetary positions in the Diary from the year –418 appear to be in zodiacal signs but cautions that the reference might nevertheless still be to zodiacal constellations, and indicates a similar ambiguity with regard to the two horoscopes from –409, cf. Hunger-Pingree (1999), pp145–146.

at all clear. According to Aristotle's pupil Eudemus of Rhodes (ca. 325 BC), quoted by Theon of Smyrna, Oenopides of Chios (ca. 450 BC) "was the first to discover the encircling belt of the zodiac...."[1] Moreover, Pliny maintains that the zodiac became known to the Greeks through Cleostratus, who lived in the latter part of the sixth century BC.[2] However, neither in Babylonia nor in Greece has any evidence for the existence of the Babylonian sidereal zodiac already in the sixth century BC been found. Yet there is the aforementioned reference by Columella, writing in the first century AD, indicating that the Babylonian zodiac had been transmitted to the Greek astronomer Meton of Athens by 432 BC, if this reference is accurate. This reference implies that Meton (and also his colleague Euctemon) observed the summer solstice in 432 BC and thus determined the location of the vernal point at that time to be at 8° Aries in the sidereal zodiac, a norm known from System B of Babylonian astronomy discussed further below. If this reference can be trusted, Meton (and presumably also Euctemon) had assimilated the Babylonian sidereal zodiac already at that early date and was able to confirm its accuracy by way of his own observation.

One route for the transmission of the sidereal zodiac from Babylon to Greece was via astrology. Thus it is known that after Alexander the Great's conquest of Babylon in −330 astrology came to Greece through figures such as the 'Chaldean' Berossus, a Babylonian astrologer who was active around −300. Berossus founded a school on the island of Cos and introduced the Greeks to astrology.[3]

As is evident from the Babylonian horoscopes found so far, the equal length (30°) fixed-star zodiacal signs of the Babylonians were used not only in the development of their astronomy but also in their astrology. Various ancient cultures practiced astrology, and the earliest known horoscopes (from the late fifth century BC) are of Babylonian origin with the planetary positions given in these horoscopes computed in relation to the zodiacal signs of the sidereal zodiac.[4]

About one century after the appearance of astrology in Babylon, a highly developed mathematical astronomy emerged there. Of special interest in consideration of the history of the sidereal zodiac is that in the Babylonian mathematical astronomy of the Seleucid era there existed two different astronomical theories to account for the movement of the moon. These are called 'System A'

---

1. Theon of Smyrna, *Mathematical Knowledge Useful for Reading Plato* III, 40.
2. Pliny, NH II, 31 (trsl. Rackham I, pp 188–189).
3. Schnabel, *Berossos und die babylonisch-hellenistische Literatur* (1923), pp 250–275, gathers the Greek fragments of the writings of Berossus.
4. Sachs (1952, 1); cf. also Rochberg (1989, 1998, 1999).

and 'System B'.[1] Both use the zodiacal signs of the sidereal zodiac but give different locations for the vernal point in the sign of Aries.

Both System A and System B of the Babylonian lunar theory utilize the location of the vernal point in the sidereal zodiac. System A places the vernal point at 10° in the sign of Aries, whilst in System B, "in contrast to System A, the vernal equinox is assumed to fall at 8° Aries instead of 10° Aries."[2] It has been suggested that System A is older than system B, and that System A became revised when it was discovered that the vernal point had shifted by 2° in the sign of Aries. This idea was put forward by Schnabel in 1923[3] and 1927[4] and led to the "frequently quoted statement that the Babylonian astronomer Kidinnu was the discoverer of the precession of the equinoxes and that this even can be dated in 379 BC, thus antedating Hipparchus by about two and one-half centuries."[5] Kidinnu, or Kidin, is the name of a Babylonian scribe associated with System B and the name Nabu-rimannu is found on a text belonging to System A, hence "it has become customary to consider Nabu-rimannu as the founder of the older system, Kidin as the founder of the younger."[6] If this is correct, then Kidinnu might be considered to have discovered that the vernal equinox had retrogressed by 2° since the time of his predecessor Nabu-rimannu, and to have modified the lunar theory accordingly.

This cannot be ascertained with any degree of certainty. It presupposes that System A originated when the vernal point was located at 10° Aries and that System B originated when the vernal point was at 8° Aries. However, it is reasonable to assume that both systems developed over an extended period of time. It also presupposes a consistent definition of the signs of the sidereal zodiac, in particular the sign of Aries, so that 8° Aries and 10° Aries refer to locations in the sidereal zodiac that can be explicitly identified relative to fixed stars belonging to the constellation of Aries. This is not necessarily the case, since the difference in degrees of the vernal point in Aries might also be on account of a difference in the definition of the starting point of the sidereal sign of Aries between System A and System B. One reason for considering this possibility is that nearly all the cuneiform texts for System A come from Babylon, whereas the majority of those for System B come from Uruk.[7] Moreover, texts for both systems are contemporaneous. This suggests that there were two schools of astronomy: one based in Babylon using a lunar theory with the

1. Neugebauer, ACT I, p41.
2. Ibid., p72.
3. Schnabel, *Berossos und die babylonisch-hellenistische Literatur*, pp227–237.
4. Schnabel (1927).
5. Neugebauer (1950).
6. Neugebauer, ACT I, p16.
7. Ibid., p42.

vernal point at 10° Aries, and the other in Uruk, whose theory placed the vernal point at 8° Aries. Is there a case for the priority of System A over System B?

> I am not even convinced that the commonly accepted priority of System A with respect to System B can be considered certain. . . . The theories as we have them before us . . . could have been developed within the relatively short period of a century or less, producing almost simultaneously several competing methods. All this is mere speculation but it seems necessary to me to express a warning against an overconfident acceptance of the traditional picture of a well-established direction of development from System A to System B.[1]

Since Schnabel's primary argument for the Babylonian discovery of precession is that System B developed from System A, his theory is undermined by the evidence on the one hand of the contemporaneity of the two systems and on the other hand of their different provinces, namely Babylon and Uruk. Whereas we know that Hipparchus deduced by way of observations that the equinoxes had shifted by about 2°, all that can be deduced from these cuneiform sources is that there existed two different locations for the vernal equinox, 2° apart. We do not know that Kidinnu or any other Babylonian astronomer observed that the equinoxes had retrogressed through the zodiac. Even if such an observation had been made, it is a large step to deduce the existence of the phenomenon of precession and to express this as a scientific theory, as Hipparchus did.

On the other hand it cannot be said with any degree of certainty that Kidinnu or any other Babylonian astronomer did not observe the phenomenon of precession, since there are three probability arguments for assigning priority to System A. Firstly, the location of the vernal equinox at 10° Aries is correct for ca. −500, whereas 8° Aries is correct for ca. −360.[2] Secondly, System B is more accurate than System A; it yields a better approximation to reality. Being a more refined astronomical system, it is not unlikely that System B was developed at a later date. The third probability argument was formulated by van der Waerden as follows:

> All our sources indicate that about 400 BC or even earlier the summer solstice was known to the Babylonians within 1 or 2 days. If System A were invented after 400 BC, the error in the equinoxes and solstices would be 2 days at least, which is improbable. About 450 BC the error would be only 1 day; this is tolerable. Probably System A was invented between 610 and 450 BC.[3]

---

1. Ibid.
2. Based on a rate of precession of 1° in 72 years, taking AD 220 as the year in which the vernal point was located at ♈ 0° and presupposing a consistent definition of the signs of the Babylonian sidereal zodiac (Appendix I).
3. van der Waerden (1968), p72.

The same probability argument can be applied to the date of invention of System B, which would place it between 465 and 300 BC (simply adding approximately 144 years to the above dates, as the shift of 2° of the vernal point from 10° Aries to 8° Aries took place over a period of ca. 144 years).

In addition, the aforementioned earliest text referring to zodiacal signs, BM 36559 + 36941 and BM 36737 + 47912 dating from the years −474 to −456, also happens to be the earliest known text to use either system and it is a System A text.[1] This would favour the priority of System A. Nevertheless, as noted above, there is not necessarily a straightforward line of development from System A to System B. What seems to be a likely possibility is that the value 10° Aries for the location of the vernal point was corrected to 8° Aries in order to give better results for the calculation of the moon's position. This cannot be thought of as the "discovery" of precession in the sense of an observation or deduction of the fact of the motion of the equinoxes. It may be concluded, therefore, in the sense that not only did he observe the phenomenon but also wrote a scientific treatise on it, thus demonstrating he had a conscious grasp of the phenomenon, that Hipparchus may be considered as the discoverer of precession.

PRECESSION IN RELATION TO THE SIDEREAL ZODIAC

Now it is possible to consider the implications for this thesis on the original definition of the zodiac of the locations of the vernal point at 10° Aries and 8° Aries in the Babylonian sidereal zodiac in Systems A and B. Commenting on this, van der Waerden writes:

> In 1963 I made an attempt to estimate the accuracy of ancient Babylonian observations of the equinoxes and solstices.... I found that about 400 BC or even earlier the summer solstice was known to within one or two days.... Kugler also investigated Tablet ACT 60 (old signature SH 93), which belongs to System A and in which the spring equinox was assumed at 10° of Aries.... His conclusion was: "An analogous calculation for Tablet No. 93 would bring us back to 500 BC ± several years."[2]

Based on an exact reconstruction of the Babylonian sidereal zodiac in Appendix I, it is evident that the vernal point was at 0° Aries in the sidereal zodiac in AD 220, and since it moves retrograde through the sidereal zodiac at

---

1. Aaboe-Sachs (1969). Britton (1999), p188, qualifies it as a "solar eclipse text with System A lunar functions," i.e., it is a text displaying some features of System A at an early stage of development.
2. van der Waerden, SA II, p266.

a rate of approximately one degree in 72 years, it is easy to see that the vernal point was located at 10° Aries around 500 BC, which is 720 years (10 x 72) prior to AD 220, confirming Kugler's statement quoted by van der Waerden above.

Van der Waerden also concluded that the Greek astronomers Meton and Euctemon drew their knowledge of the zodiac from the Babylonians, adopting System B:

> These two astronomers observed the summer solstice in Athens in 432 BC (Ptolemy, *Almagest* III, 1).... Meton placed the equinoxes and solstices at 8 degrees Aries, 8 degrees Cancer, etc. (Columella, *De re rustica* IX, 14), exactly like the Babylonian System B calculations.[1]

According to the reconstruction of the sidereal zodiac in Appendix I the vernal point was located at 8° Aries actually around 356 BC—144 years (2 x 72) after 500 BC—since it moved two degrees from 10° to 8° Aries in 144 years. Therefore, based on the reconstructed sidereal zodiac in Appendix I, it is evident that Meton was approximately one degree in error (a negligible amount for early Greek astronomy in the fifth century BC), since the vernal point was at 9° Aries in about 428 BC, midway between 500 BC and 356 BC. All of these considerations depend, of course, upon a consistent definition of the Babylonian sidereal zodiac such as that in Appendix I. Only in the case of a consistent definition is it possible to arrive at reliable dates for the location of the vernal point at specific degrees of the sidereal zodiac.

With regard to the question of a consistent definition of the sidereal zodiac, Huber's analysis of a number of Babylonian astronomical observations and computations (reduced to the epoch −100) leads to the conclusion that in −100 the vernal point was located at 4°28′ in the sidereal zodiac, and this finding is completely consistent with the reconstructed sidereal zodiac in Appendix I.[2]

Thus the theoretical considerations presented in this thesis leading to an *intrinsic definition* of the Babylonian sidereal zodiac and—independent of these theoretical considerations—also the astronomical observations and computations of the Babylonians themselves, according to Huber's analysis, reveal an inner consistency in the definition of this zodiac.

Even without a consistent definition of the Babylonian sidereal zodiac, it might still be possible to draw conclusions regarding the history of the zodiac

---

1. Ibid., p246.
2. Huber (1958). The reconstructed Babylonian sidereal zodiac in Appendix I agrees exactly with Peter Huber's analysis of Babylonian astronomical observations for the epoch -100, according to which he located the vernal point at 4°28′ Aries in the sidereal zodiac, since during the 320 years from −100 to AD 220 (the date specified in Appendix I when the vernal point was located at 0° Aries) the vernal point retrogressed 4°27′ to reach 0° Aries, allowing a rate of precession of one degree in 72 years ($320/72 = 4.444° = 4°27′$).

from a consideration of the regression of the vernal point through the signs of the zodiac, particularly in relation to the retrogression of the vernal point through the sign of Aries. However, in order to continue with these considerations, it is helpful to proceed under the assumption that there *was* a consistent definition of the sidereal zodiac, which was also the assumption underlying Huber's analysis. According to the reconstructed Babylonian sidereal zodiac (Appendix I), a frame of reference for the precession of the vernal point through the latter part of the sign of Aries from 10° Aries in –500 to 0° Aries in AD 220 is given.

As a starting point for further observations concerning the retrogression of the vernal point through the sidereal zodiac during this time, let us consider Goold's conjecture based on the assumption that at the time of Hipparchus in the second century BC the vernal point was at 0° Aries:

> It is the great glory of Hipparchus to have discovered the precessional shift, apparently by comparing his observations of Spica with those of Timocharis. . . . Eudoxus, we are told (Hipparchus 2.1.18) placed the vernal point at 15 degrees Aries, Aratus at the beginning of the sign [0 degrees Aries]. Now since their chronological difference corresponds to a precessional shift of not much more than one degree, it is obvious that both are struggling to preserve conventions that do not fit the phenomena. Occasional references to 8 degrees [Aries] may be related, as Neugebauer suggests, to the vernal point of System B of the Babylonian lunar theory; and 8 degrees [Aries] or thereabouts may well have marked the [vernal] equinox when the zodiac as we know it was devised (the Romans—Caesar, Vitruvius, Columella, Pliny—generally adopted 8 degrees [Aries]). . . . If in the time of Hipparchus the vernal equinox occurred at the first point of Aries [0 degrees Aries], then in the time of Ptolemy it must have occurred at about 26 degrees Pisces, and today it must occur at about 1 degree Pisces.[1]

If Hipparchus did believe that at his time the vernal point was at 0° Aries, then—like Meton—he was in error, because in terms of a consistent definition of the sidereal zodiac the vernal point did not reach 0° Aries until AD 220.[2] Based on the reconstructed Babylonian sidereal zodiac (Appendix I), at the time of Hipparchus (around 140 BC) the vernal point was actually at 5° Aries, since it shifted five degrees in the 360 years (5 × 72) between 140 BC and AD 220. Nevertheless, according to Goold, there is evidence that Hipparchus believed that the vernal point was located at 0° Aries at his time, which seems to indicate the possibility that Hipparchus was responsible for introducing the tropical zodiac into astronomy by simply assuming that the vernal point *in his*

---

1. Manilius, *Astronomica*, introduction by G.P. Goold, pp LXXXII–LXXXIII. Additions in brackets [ ] by R.P.
2. Appendix I.

day was located at 0° Aries, just as Meton from his observation of the summer solstice at 8° Cancer in 432 BC must have assumed that the vernal point was at 8° Aries some three hundred years before Hipparchus. On the other hand, since Hipparchus discovered the movement of the vernal point ("precession of the equinoxes") in relation to the fixed stars, it is quite reasonable to conclude that *precisely on this account* he decided to take the vernal point as his point of reference for all measurements rather that choosing a sidereal frame of reference as the Babylonians had done.

> 10° Aries and 8° Aries were the Babylonian longitudes for the equinoxes and particularly the latter norm also found widespread acceptance in the west. Hipparchus informs us that Eudoxus placed the midpoints (15°) of the signs at these points whereas he himself, following "most of the old mathematicians" (and Aratus) reckoned the seasons from the beginning of the signs.[1]

Nearly three hundred years after Hipparchus, Ptolemy also adhered to the vernal point as the fixed point of reference for all his measurements. Although, just as in consideration of Hipparchus, the possibility exists that he simply assumed that the vernal point was at 0° Aries *in his day* (he wrote the *Tetrabiblos* around the year AD 150, some seventy years before AD 220), when it was actually at 1° Aries in terms of a consistent definition of the Babylonian sidereal zodiac (Appendix I). Ptolemy stated in the *Tetrabiblos*: "It is reasonable to reckon the beginnings of the signs also from the equinoxes and solstices."[2] Ptolemy's error in ca. AD 150 in placing the vernal point at 0° Aries (rather than at 1° Aries, where it was actually located in the sidereal zodiac at his time—see Appendix I) amounts to only one degree, the same amount as Meton's error in 432 BC when he placed the vernal point at 8° Aries instead of at 9° Aries where it was located at his time. These 'errors' are real but would not have been apparent to either Meton or Ptolemy, since it is solely from the vantage point of the reconstructed Babylonian sidereal zodiac (Appendix I) that it is possible (now, i.e. retrospectively) to make an exact specification of the location of the vernal point at different historical dates.[3]

In contrast, the apparent error made by Hipparchus in placing the vernal

---

1. Neugebauer, HAMA I, p278.
2. Ptolemy, *Tetrabiblos* I, 22 (trsl. Robbins, p109). This formulation by Ptolemy could be interpreted that he acknowledged the possibility for other definitions of the beginnings of the signs, i.e. other zodiacal frameworks not based on the vernal point at 0° Aries.
3. As discussed above, it does not seem likely that the Babylonians had a conscious grasp of the phenomenon of precession, let alone any knowledge of the precise rate of precession. In this case they would not have been able to compute the location of the vernal point in the sidereal zodiac accurately at different points in time. This is only possible now, retrospectively, on the basis of a grasp of all the factors involved (including knowledge of the exact rate of precession).

point at 0° Aries in his day was much larger, amounting to five degrees, since the actual location of the vernal point in –140 was 5° Aries (Appendix I). Let us examine this apparent error on the part of Hipparchus more closely. According to his own testimony, Hipparchus knew of different statements concerning the location of the vernal point:

(1) at 15° Aries by the astronomer and mathematician Eudoxus (fourth century BC); and

(2) at 0° Aries by the poet Aratus (third century BC), where it should be borne in mind that Aratus was probably referring to the calendrical system of Euctemon rather than to an astronomical coordinate system and that it was probably an interpretation on Hipparchus' part, as an astronomer, that this signified a location of the vernal point at 0° Aries when he heard that Aratus "reckoned the seasons from the beginning of the signs."

Since Hipparchus knew about the precession of the equinoxes (retrogression of the vernal point), and had estimated this movement to be not more than 1° per century, it must have been clear to him that Eudoxus and Aratus, who lived less than a century apart, could not both have been right. Also, Hipparchus utilized some Babylonian parameters in his astronomical work.[1] So it is possible that he knew about the Babylonian norms for the vernal point at 10° Aries and 8° Aries in the sidereal zodiac. Nevertheless, probably for practical reasons he chose to make his measurements in relation to the equinoctial and solstitial points rather than in the sidereal framework offered by the Babylonian zodiac, and it was precisely this approach that enabled him to discover the precession of the equinoxes. Indeed, as the discussion below indicates, Hipparchus, by equating the equinoxes and solstices with the beginnings of the signs, thus setting the vernal point at 0° Aries, effectively introduced the ecliptic coordinate system of the tropical zodiac into astronomy. The evidence that Hipparchus introduced the ecliptic coordinate system of the tropical zodiac into astronomy is inconclusive, since in his only extant work *Commentary to Aratus* he does not use ecliptic coordinates. The most important evidence is provided by Ptolemy *Almagest* VII, 2, who indicates that Hipparchus found the star Regulus to have been 29°50' east of the summer solstitial point and that he observed the star Spica to have been 6° west of the autumnal equinox in his (Hipparchus') time.[2]

In Chapter 3 the tropical zodiac is discussed as a spatial projection of the tropical calendar (see Figures 9 and 10). It is important, therefore, to distinguish between the *tropical calendar* of Euctemon comprising a division of the

1. Toomer (1980).
2. Ptolemy, *Almagest* VII, 2; cf. Evans (1998), p 260.

cycle of the year into twelve solar months and the *tropical zodiac* of Hipparchus. For Euctemon the solstices and equinoxes were *moments in time* from which to begin the counting of the days of the solar months. For Hipparchus the solstices and equinoxes were projected onto the celestial sphere and used as spatial locations from which to make measurements, rather than using the zodiacal fixed stars as the frame of reference for such measurements as was the Babylonian practice. Possibly because Euctemon named the solar months after the twelve signs or constellations of the zodiac, Hipparchus called the spatial divisions arising as 30° arcs measured from the solstices and equinoxes *zodiacal signs*. Thus, according to Ptolemy, Hipparchus referred to Spica as "about 22° Virgo in the time of Timocharis,"[1] i.e. this was Hipparchus' way of expressing that Spica was 8° west of the autumn equinoctial point[2] (derived from an observation of Timocharis around 284 BC — Figure 8). And then Hipparchus, based on his own observation around 135 BC, concluded that at his time Spica, since he had observed that this star was then only 6° west of the autumnal point (Figure 8), was about 24° Virgo.[3]

Here it should be noted that Hipparchus' statement does not necessarily imply that Timocharis also used the tropical zodiac as his frame of reference. All that can be concluded is that Timocharis had made an observation yielding the distance of Spica 8° westward of the autumnal point and either he or Hipparchus (or later Ptolemy) translated this into the longitude of 22° Virgo, which is one of the earliest known instance of the use of longitudes expressed in terms of signs of the tropical zodiac, more commonly called the *ecliptic coordinate system*.[4]

> Who was the inventor of the method of determining the positions of stars according to the ecliptic coordinate system? Until now it was assumed to be fairly certain that Hipparchus—in his star catalogue—was the first to use this method of determining positions.... Already long before Hipparchus astronomers used

---

1. Ptolemy, *Almagest* VII, 2; cf. Neugebauer, HAMA I, p287, n30.

2. The possibility exists that Hipparchus did not actually use the formulation *about 22° Virgo* but that he had simply referred to Spica being 8° west of the autumnal point and that it was Ptolemy who translated this indication into the formulation *about 22° Virgo*.

3. This was one of only two ecliptic longitudes for fixed stars (the other being for Regulus) that Hipparchus supposedly recorded in his star catalogue; cf. Neugebauer, HAMA I, p283, n13. The Hipparchian star catalogue itself is no longer extant, but since a substantial part of it is embedded in Hipparchus' *Commentary to Aratus*, it is possible to derive knowledge of the Hipparchian star catalogue from this source; cf. Vogt (1925).

4. The only previously known specifically Greek astronomical observation given in terms of ecliptic coordinates is the famous one quoted above concerning the observation of the summer solstice by Meton of Athens in 432 BC locating the summer solstitial point at 8° Cancer and thus the vernal point at 8° Aries, which clearly applies to the Babylonian sidereal zodiac and not to the tropical zodiac. Cf. van der Waerden, SA II, p246.

the degrees of the zodiacal signs for their indications of longitude.... I am thinking of Timocharis and Aristyllos or their students. In their star lists, of which we have little knowledge, they must have already used this norm for determining longitudes. Following their example Hipparchus then applied this method and made corrections through exact observation for his time.[1]

As Gundel points out, we have little knowledge of the star lists of Timocharis or his associate Aristyllos, both of whom made their stellar observations in Alexandria in the third century BC. Apart from the one observation relating to Spica utilized by Hipparchus, there is only a list of declinations of eighteen stars measured by Aristyllos and Timocharis, reported by Ptolemy in the *Almagest*.[2] These astronomical observations do not indicate that Timocharis and Aristyllos used the ecliptic coordinate system of the tropical zodiac as postulated by Gundel.

Further, the study by Vogt of Hipparchus' star catalogue reveals that primarily he employed *polar longitude*, a coordinate system used also in Indian astronomy.[3] According to Ptolemy's indications in *Almagest* VII, 2 referred to above, the zodiacal stars Regulus (α Leonis) and Spica (α Virginis) were evidently the stars used by Hipparchus for determining precession. In the case of Spica, as referred to already, Hipparchus determined that Spica was 6° west of the autumnal point, i.e. at 24° Virgo in the tropical zodiac. Regarding Regulus, "According to Ptolemy, Hipparchus observed this star in the 50th year of the third Callippic period (129/128 BC) and found it to be 29°50' east of the summer solstitial point."[4] Translated into longitude in the tropical zodiac, this means that Hipparchus observed Regulus at that time to be at 29°50' Cancer.[5]

Here the difference between the new system used by Hipparchus and the Babylonian method of recording observations in the sidereal zodiac becomes apparent. In the reconstructed Babylonian sidereal zodiac (Appendix I) Regulus' longitude is 5° Leo, Spica's is 29° Virgo. A 5° shift is apparent: for Spica, from 29° Virgo (Babylonian sidereal) to 24° Virgo (Greek tropical); and for Regulus, from 5° Leo (Babylonian sidereal) to 29°50' Cancer (Greek tropical).

---

1. Gundel, *Neue astrologische Texte des Hermes Trismegistos*, p134.
2. Ptolemy, *Almagest* VII, 3; cf. Goldstein and Bowen (1989).
3. Vogt (1925); cf. also Neugebauer, HAMA I, p279. The *polar longitude* of a star S is the angular distance from the zero point (0° Aries) to the intersection with the great circle connecting the celestial poles, i.e. of the arc NS with the ecliptic, where N is the north celestial pole. In Indian astronomy 0° Aries is the zero point of the sidereal zodiac. For Hipparchus the zero point (0° Aries) was identified with the vernal point.
4. Evans (1998), p260.
5. This is the only other example (apart from Spica) of Hipparchus assigning a star a longitude in the ecliptic coordinate system of the tropical zodiac. The majority of stellar longitudes given in his *Commentary to Aratus* are *polar longitudes* (n73); cf. Vogt (1925).

This 5° shift is due to the location of the vernal point at the time of Hipparchus at 5° Aries. For the history of the zodiac, therefore, it can be concluded that just as the Babylonians were the inventors of the sidereal zodiac, Greek astronomers (probably Hipparchus, some 265 years prior to Ptolemy) invented the tropical zodiac.

Thus, returning to Gundel's question: Who was the inventor of the method of determining the positions of stars according to the tropical ecliptic coordinate system? The answer cannot be given with any degree of certainty, but since Hipparchus—according to Ptolemy's testimony—was the first astronomer who is known to have definitely used this system, even if only for the two stars Regulus and Spica, and since it was adopted also by Ptolemy and continues to be used to the present day, it can be stated that Hipparchus *effectively* introduced the tropical zodiac into astronomy.

One could conjecture that it was prior to his discovery of the precession of the equinoxes that Hipparchus wrote the *Commentary to Aratus*, in which he used primarily polar longitudes, and that after this discovery—having recognized the significance of the ecliptic—he used the ecliptic coordinate system of the tropical zodiac in compiling his catalogue of stars, now no longer extant.[1] The reasoning underlying this conjecture is precisely the line of argument used by Ptolemy in compiling his star catalogue (*Almagest* VII and VIII), where he gives star positions "in terms of longitude and latitude, not with respect to the equator, but with respect to the ecliptic, [i.e.] as determined by the great circle drawn through the poles of the ecliptic and each individual star."[2]

The question still remains whether the tropical zodiac came into existence because Hipparchus simply assumed that at his time the vernal point was at 0° Aries. From Hipparchus' own statement referred to above, he knew that his predecessor Eudoxus had located the vernal point at 15° Aries. This is clearly in terms of the sidereal zodiac using a norm, however, that was otherwise previously unknown (the known Babylonian ones being 8° Aries and 10° Aries). Hipparchus also knew that Aratus "reckoned the seasons from the beginning of the signs."[3] As indicated above, this statement could refer to the tropical

---

1. Neugebauer, *The Exact Sciences in Antiquity*, p 69: "Of Hipparchus' writings, only his *Commentary to Aratus* is preserved.... This work is undoubtedly an early work of Hipparchus, written before the discover of precession. This follows from the fact that the positions of stars are never given in ecliptic coordinates (longitude and latitude) but in a mixed ecliptic-declination system.... It was obviously the discovery of precession that later led Hipparchus to introduce real ecliptic coordinates because longitudes increase proportionally with time whereas latitudes remain unchanged."
2. Ptolemy, *Almagest* VII, 4 (trsl. Toomer, p 339).
3. Neugebauer, HAMA I, p 278.

calendar of Euctemon in which the beginning of the seasons is specified by the equinoxes and solstices. Evidently for practical purposes, since presumably he was confident that he was able to determine the positions of the equinoctial and solstitial points relatively accurately, Hipparchus chose to use these points as his frame of reference in making astronomical observations, thus departing from the Babylonian practice of making observations in relation to the background of the fixed stars and determining the positions of the equinoctial and solstitial points against that background. It is reasonable to conclude that Hipparchus' assumed adoption of the ecliptic coordinate system of the tropical zodiac was suggested to him when he heard of the above statement by Aratus who, in turn, perhaps had gained this idea by way of transmission ultimately from Euctemon. Seen in this light, if the sequence of transmission (Euctemon—Aratus—Hipparchus) is correct, the tropical zodiac came into existence effectively as a spatial projection on the part of Hipparchus of Euctemon's tropical calendar (Chapter 3), even if Hipparchus himself did not conceive of the tropical zodiac in this way.

An alternative possibility, implicit in the above quotation from Goold, is that Hipparchus simply assumed that *at his time* the vernal point was at 0° Aries. This alternative finds support in the statement of Columella (first century AD): "I am well acquainted with the reckoning of Hipparchus, which declares that the solstices and equinoxes occur not in the eighth but in the first degrees of the signs of the zodiac."[1] Columella's remark seems to imply that Hipparchus relocated the vernal point from the Babylonian norm (System B) of 8° Aries to 0° Aries, which then led to the use of the tropical zodiac in Greek astronomy as a basic astronomical coordinate system for measuring the positions of the sun, moon, planets, and fixed stars. In this case it could be said that the sidereal zodiac of the Babylonians literally gave birth to the tropical zodiac of the Greeks by way of it being fixed permanently with the vernal point at 0° Aries.

Returning to the above quotation from Goold, if it were to be formulated giving the actual positions[2] of the vernal point at that time and in the present, it would read: "At the time of Hipparchus the vernal equinox occurred at 5 degrees Aries, then in the time of Ptolemy it must have occurred at about 1 degree Aries; and today it must occur at about 5 degrees Pisces." Continuing with the words of Goold:

> Today in fact the effect of precession has been to move every zodiacal sign twenty-nine degrees away from where, according to astrological doctrine, it

---

1. Columella, *De re rustica* IX, 14 (trsl. Ash-Forster-Heffner II, pp 487–489).
2. In terms of the reconstructed Babylonian sidereal zodiac (Appendix I).

ought to be. Oddly enough it is Ptolemy who has saved the day for astrologers. In *Tetrabiblos* I, 22 the astronomer virtually says that for astronomical purposes he will define the first point of Aries as the vernal equinox. If that moves, then the whole zodiac will just have to move with it. For astrological purposes men had better look to this movable, artificial zodiac. And so it has come to pass. When today's readers of almanacs are informed that the Sun travels through Aries from March 21 to April 20, the name Aries denotes not the group of stars so identified and marked in our star atlases, but thirty degrees of the ecliptic measured off from the vernal equinox, a length of line constantly moving and today almost entirely contained in the astronomical constellation of Pisces.[1]

In terms of the reconstructed Babylonian sidereal zodiac in Appendix I, the vernal point has not yet shifted back 29 degrees as stated here by Goold, but rather it has shifted back approximately 25 degrees from 0° Aries to its current location at about 5° Pisces.[2] Nevertheless, Goold's statement still holds true, provided we modify 29 degrees to 25 degrees. If we follow Goold's interpretation, confirmed by Hipparchus' own statement quoted above, there were various systems locating the vernal point at different degrees in the sidereal sign of Aries, and Hipparchus' system—locating the vernal point at 0° Aries—was one of them. According to Goold this signified the creation of a "movable, artificial zodiac," i.e. the *tropical zodiac*. Regardless of Goold's interpretation, it is reasonable to assume that Hipparchus, in creating the tropical zodiac, was following the model of the Babylonian sidereal zodiac, i.e. twelve signs each 30° long, and thus it is possible to speak of the influence of Babylonian astronomy upon this new definition of the zodiac.

Whether or not we accept Goold's interpretation, it is clear that Hipparchus and Ptolemy changed things completely from the approach of Babylonian astronomy. For Meton and Eudoxus, for example, it was evidently natural to ask what degree the summer solstitial point or vernal point was located in the (sidereal) zodiac in their day. *Sidereal* is written here in parentheses, because as long as the Babylonian sidereal zodiac was the *only* zodiac, there was no need to specify it as the *sidereal* zodiac; it was simply *the* zodiac. By fixing the vernal point at 0° Aries, the Babylonian sidereal zodiac became replaced by the tropical zodiac which, as discussed above, was a new creation on the part of Greek astronomy, something to which evidently the Babylonian zodiac gave birth. Thus in the period after Ptolemy—at least in the west—the sidereal zodiac of the Babylonians was completely forgotten, until it was brought to

---

1. Manilius, *Astronomica*, introduction by G. P. Goold, pp LXXXIII–LXXXIV.
2. According to the reconstructed Babylonian zodiac in Appendix I, the vernal point was at 0° Aries in AD 220. From AD 220 to AD 2000 is a period of 1780 years and, allowing a rate of precession of 1° in 72 years, the shift back from 0° Aries amounts to almost 25 degrees, since 1780/72 = 24.72°.

the light of day again through the excavation and decipherment of cuneiform texts during the nineteenth and twentieth centuries. The central focus of this thesis on the definition of the zodiac is to draw attention again to the *original* zodiac—the Babylonian sidereal zodiac—which in the sense described above possibly gave birth to the tropical zodiac by way of the fixing of the vernal point once and for all at 0° Aries. It remains an open question concerning the extent of conscious knowledge on the part of Hipparchus or Ptolemy as to the consequences of this fixing of the vernal point at 0° Aries. Or did each of them simply believe that the vernal point was located at 0° Aries at the time they were living? Since each of them knew of precession, it must have been clear to them that the vernal point could not remain fixed at 0° Aries permanently.

The most reasonable conclusion is that it was purely practical astronomical considerations that led them to work with the ecliptic coordinate system of the tropical zodiac. For, without having knowledge of a precise definition of the Babylonian sidereal zodiac (such as that in Appendix I), i.e. a definition which would have enabled them to know the precise location of the equinoctial and solstitial points in the sidereal zodiac (rather than the imprecise indications of Meton and Eudoxus that the vernal point lay at 8° Aries or at 15° Aries), they had no alternative but to record their observations in the framework provided by the equinoctial and solstitial points.

Another obstacle in the way for the Greek astronomers to work with the Babylonian sidereal zodiac was that, without knowing the exact coefficient of precession (ca. 1° in 72 years), even if they had had knowledge of a precise definition of the sidereal zodiac, they would have been unable to interface effectively between their observations made in relation to the equinoctial and solstitial points and the coordinate system of the sidereal zodiac. It is only through achievements such as those attained by modern exact astronomy that this possibility is readily and easily given.

In conclusion, therefore, it was probably through Hipparchus that the tropical zodiac was *effectively* introduced into astronomy as a new astronomical coordinate system later used by Ptolemy in the *Almagest* as the standard system, and it was through Ptolemy that the tropical zodiac was introduced into astrology in his astrological textbook the *Tetrabiblos*, which later became the 'bible' of astrology in the Muslim world (Chapter 3).

For it is clear that prior to Ptolemy astrology was sidereal, as is evident, for example, from Goold's translation of the *Astronomica* by the Roman astrologer Manilius, who lived at the beginning of the first century AD:

Resplendent in his golden fleece the Ram leads the way and looks back with wonder at the backward rising of the Bull, who with lowered face and brow summons the Twins; these the Crab follows, the Lion the Crab, and the Virgin the Lion. Then the Balance, having matched daylight with the length of night, draws on the Scorpion, ablaze with his glittering constellation, at whose tail the man with the body of a horse aims with taut bow a winged shaft, ever in act to shoot. Next comes Capricorn, curled up with his cramped asterism, and after him from urn upturned the Waterman pours forth the wanted stream for the Fishes which swim eagerly into it; and these as they bring up the rear of the signs are joined by the Ram (*Astronomica* I, 263–274).... These then are the constellations which decorate the sky with even spread (*Astronomica* I, 532).... How great is the space occupied by the vault of the heavens and how great the territory within which the twelve signs of the zodiac move (*Astronomica* I, 539)... distribute thirty degrees to each sign (*Astronomica* III, 493).... Some ascribe these powers [the equinoxes and solstices] to the eighth degree; some hold that they belong to the tenth; nor was an authority lacking to give to the first degree the decisive influence and the control of the days (*Astronomica* III, 680–682).[1]

Manilius, in this most ancient complete surviving Latin astrological text, is clearly referring here to the 30° signs of the sidereal zodiac, indicating various possible locations of the vernal point in the sidereal zodiac. Goold comments:

Whereas Manilius mostly locates the actual tropic degree [of the equinoxes and solstices] at the beginning of a tropic sign, here he places it within the sign and specifically in the eighth or tenth degree. Indeed, his last line, referring to the authority who assigned the tropic to the first degree, even suggests eccentricity on that person's part.[2]

Was this authority Hipparchus? As referred to already, according to the testimony of Columella (first century AD), it was indeed Hipparchus who placed the solstices and equinoxes "in the first degrees of the signs of the zodiac."[3] In any case it is clear that Manilius was not using the tropical zodiac as a *de facto* system since he refers to System A (vernal point at 10° Aries) and System B (vernal point at 8° Aries), as well as to Hipparchus' system (vernal point at 0° Aries). The whole text quoted above from the *Astronomica* by Manilius demonstrates that the entire work is *sidereal*, and the following sentence indicates that his definition of the sidereal zodiac was very close to the original Babylonian definition: "As he emerges in his backwards rising with head hanging down, the Bull brings forth in his sixth degree the Pleiades" (*Astronomica* V, 140–142).[4] In the reconstructed Babylonian sidereal zodiac (Appendix I) the Pleiades are located at 5° Taurus. Therefore there was at most a one degree

---

1. Manilius, *Astronomica* (trsl. G.P. Goold), pp 25–27, 47, 219. Comment in brackets [ ] added by R.P.
2. Manilius, *Astronomica*, introduction by G.P. Goold, p LXXXI.
3. Columella, *De re rustica* IX, 14 (trsl. Ash-Forster-Heffner II, pp 487–489).
4. Manilius, *Astronomica* (trsl. G.P Goold), p 311.

difference between Manilius' assignment of a longitude to the Pleiades in Taurus and the Babylonian specification of a longitude to this stellar cluster in the sidereal zodiac, and more likely there was no difference at all, since the expression 'sixth degree' was probably according to the ancient method of counting degrees and thus was actually referring to 5° as a numerical value.[1] This example demonstrates the utility of the reconstructed Babylonian sidereal zodiac (Appendix I) in assessing such statements by various authors in antiquity, or for the comparison of stellar observations from ancient sources. It also shows that the Babylonian sidereal zodiac was transmitted—via astrology—to Rome.[2]

To summarize: the phenomenon of precession implicitly came to expression in antiquity through the custom of Greek astronomers such as Meton and Eudoxus, as well as in the practice of the Babylonian astronomers who devised System A and System B, assigning the vernal point to a particular degree in the sidereal zodiac. The very existence of this custom in antiquity presupposes the existence of the sidereal zodiac. This custom was later extrapolated by Newton, who believed that thereby he had found a new method of astronomical chronology (Appendix II), although Newton did not have access to an exact definition of the sidereal zodiac and thus could not develop his method of astronomical chronology in any precise way. Now, however, since the recovery of the Babylonian sidereal zodiac through the excavation and decipherment of cuneiform texts enabling its exact reconstruction (Appendix I), it is possible to refer to the retrogression of the vernal point through the sidereal zodiac in a more precise way, i.e. associating historical dates with particular locations of the vernal point in the sidereal zodiac, assuming that the reconstruction is accurate.

---

1. Neugebauer, HAMA I, p279.
2. Cramer, *Astrology in Roman Law and Politics*, p169 mentions that Hadrian (second century AD) became emperor, according to an astrological prognosis, because in his horoscope "the moon too was about to be in conjunction with a bright fixed star in the twentieth degree. For one must not only pay attention to the conjunction of the moon with the planets, but also with fixed stars." This is a clear statement that here the astrologer's frame of reference was the sidereal zodiac which, along with the doctrines and practices of the 'Chaldaeans', as the astrologers were generally known in the Roman empire, was transmitted to Rome just as it had been transmitted to Greece and Hellenistic Egypt. Thus the sceptic philosopher Sextus Empiricus (ca. AD 200) in his work *Adversus mathematicos* (5, 5) refers to the astrologers as Chaldaeans for whom "all these stars and constellations were considered as prime importance in each horoscope. Each zodiacal constellation . . . was subdivided into 30 degrees. . . ." (Cramer, ibid., p205). This is again a clear reference to the sidereal zodiac—presumably of the Babylonians (= Chaldaeans).

3

THE TROPICAL ZODIAC

IN CHAPTER 1 the observation is made that what is known as the zodiac (with twelve signs) in popular terminology must be clearly distinguished from the astronomical zodiac comprising twelve constellations.[1] The former was originally a calendar scheme (Euctemon's tropical calendar—Figure 2) dividing the solar year into twelve equal months commencing with the vernal equinox, in which each solar (tropical) month is named after one of the twelve signs: Aries, Taurus, Gemini, etc.[2] The latter is a division of the celestial sphere into twelve sectors in which the stars in each sector appeared to astronomers in antiquity to form certain configurations, e.g. the stars in the Aries sector appeared to form the figure of a ram, those of Taurus a bull, and so on.

Moreover, as discussed in Chapter 2, an interface between the tropical calendar and the zodiac is given by way of determining the zodiacal location of the vernal point for a given date. In this way the purely temporal phenomenon of the passage of the solar (tropical) months was placed into connection with the sun's ecliptic location specified in relation to the background of the fixed stars. Thus it was customary to determine the sun's zodiacal position on the day of the vernal equinox, at the start of the solar year. This position could

---

1. As the title of this chapter indicates, the name of that which is nowadays simply referred to as 'the zodiac' in popular terminology is—more correctly—the *tropical zodiac*. As discussed at length in this book, the tropical zodiac is to be distinguished from the *sidereal zodiac* (equal 30° zodiacal constellations) and the *astronomical zodiac* (unequal-division zodiacal constellations). The word 'tropical' obviously relates to the *tropic of Cancer* and the *tropic of Capricorn*, where the sun turns on reaching its greatest declination north or south—the maximum northerly declination of the sun being attained on the day of the summer solstice and the maximum southerly declination on the day of the winter solstice. The summer and winter solstices are the *turning points* on the sun's path, called in Greek *tropai helioio* ('turnings of the sun'). These two turning points on the sun's path, together with the vernal and autumnal equinoctial points define the solar year upon which the tropical zodiac is based.

2. Figures 3 and 4 indicate the definition of the solar (tropical) months. Euctemon of Athens (432 BC) was the originator of this solar calendar; cf. Pritchett-van der Waerden (1961).

also be ascertained by observing the sun's zodiacal location at the summer or winter solstice or at the autumn equinox.

For example, according to Columella, writing in the first century AD, the Greek astronomer Meton of Athens, who observed the summer solstice in the year 432 BC, "placed the equinoxes and solstices at 8° Aries, 8° Cancer, etc. (Columella, *De re rustica* IX, 14)...."[1] This specification was obtained by projecting the sun's position at the time of the vernal equinox, known as the vernal point, onto the fixed star background of the sidereal zodiac. Many problems were entailed here, in order for this projection to be at all accurate, not the least of which was a general lack of knowledge as to how exactly the sidereal zodiac was defined. Another problem involved here was the movement of the vernal point against the background of the sidereal zodiac, which, as far as can be ascertained, was not known prior to its discovery in the second century BC. The *precession of the equinoxes*, as the retrogression of the vernal point against the background of the sidereal zodiac is known, was discovered by Hipparchus, as discussed in Chapter 2. The slow, retrograde movement of the vernal point is expressed schematically in Figure 9, which indicates the transition of the vernal point from the sidereal division of Aries the Ram to Pisces the Fishes (that took place in ca. AD 220 according to the reconstructed Babylonian sidereal zodiac in Appendix I).

The vernal point is obtained by projecting a line from the earth through the sun up to the fixed stars on the day (actually at the moment) of the vernal equinox. Similarly, the autumnal point is defined by the projection of the earth-sun line onto the ecliptic on the day of the autumnal equinox. In the same manner the tropical (solstitial) points are obtained: the summer tropical (solstitial) point is determined by the extended earth-sun line on day 1 of the zodiacal (tropical) month of Cancer, and the winter tropical (solstitial) point is found on the ecliptic by projecting the earth-sun line on the day of the winter solstice, day 1 of Capricorn. In fact, by making such a spatial projection not only on Aries 1 (vernal equinox), Cancer 1 (summer solstice), Libra 1 (autumnal equinox) and Capricorn 1 (winter solstice) but also on the first day of each remaining solar month (Taurus 1, Gemini 1, Leo 1, Virgo 1, Scorpio 1, Sagittarius 1, Aquarius 1 and Pisces 1), a twelvefold spatial division relating to the solar months is obtained. This division is known as the *tropical zodiac*. Figure 10 shows the tropical zodiac in relation to the unequal constellations of the astronomical zodiac at the present time.

As discussed in Chapter 2, the tropical zodiac appears to have originated with Hipparchus, and it is evident that the tropical zodiac is simply a spatial projection of Euctemon's tropical calendar (Figures 2 and 10).

1. Van der Waerden SA II, p 266.

Since the time of Ptolemy the orthogonal ecliptic coordinate system of the tropical zodiac has been used by astronomers. As indicated in Chapter 2, it is customary to associate Hipparchus with the introduction of the ecliptic coordinate system of the tropical zodiac in his lost catalogue of stars, even though there is only circumstantial evidence for this. And it would seem that this assumption concerning Hipparchus' use of tropical ecliptic coordinates in his lost catalogue of stars applies only to ecliptic longitudes, not to latitudes, since no Hipparchian latitude has ever been referred to or recorded, although in *Almagest* VII, 3 Ptolemy quotes Hipparchus in referring to the constancy of the latitude of Spica (2° south) between the time of Hipparchus and himself.

Figure 9

Solar Months and Zodiacal Constellations

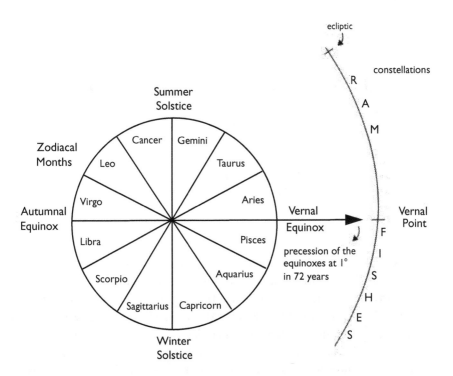

Figure 10

Solar Months and Zodiacal Constellations

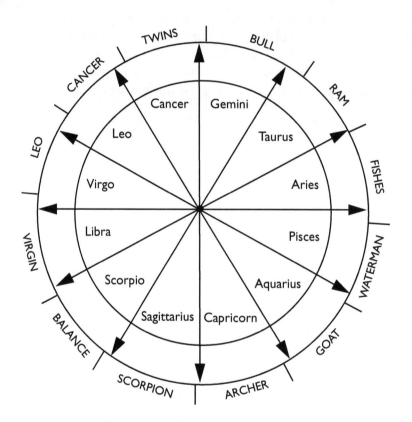

In fact, there is no *direct* evidence that anyone prior to Ptolemy utilized the orthogonal ecliptic coordinate system of the tropical zodiac to determine the locations of stars.[1] The tropical zodiac thus replaced the Babylonian sidereal zodiac, which was the forerunner of the tropical zodiac. As discussed in Chapter 2, it was evidently the Babylonian zodiac which provided the framework for Meton's measurement of the vernal point in 432 BC referred to above.

In general, following the assumed adoption of the ecliptic coordinate system of the tropical zodiac by Hipparchus, astronomical specification of the locations of stars were given in terms of the number of degrees within the twelve

---

1. Cf. Neugebauer, HAMA I, p280.

zodiacal signs along the ecliptic measured from the vernal point. One example of this is provided by the star catalogue in the astrological treatise *Liber Hermetis Trismegisti* embedded in a Latin manuscript from the fifteenth century that was unexpectedly found in the British Museum Library in the 1930's in which the longitudes of the stars are given as tropical longitudes in zodiacal signs.[1] The longitudes indicate that this star catalogue was compiled around the time of Hipparchus or shortly thereafter,[2] and the terminology used to describe the stars is Hipparchian in nature, reminding us of the descriptions used in the *Commentary to Aratus*. The prime example, however, is Ptolemy's catalogue of stars from Books VII and VIII of the *Almagest*, in which the longitudes of stars are given as orthogonal ecliptic coordinates in the tropical zodiac (Appendix I). As discussed below, Arabic astrologers—following the introduction by Ptolemy of the tropical zodiac into astrology—also used the tropical zodiac, employing various methods of astronomical reckoning to find the positions of the planets in the signs of the tropical zodiac.[3]

Thus two different uses of the ecliptic coordinate system of the tropical zodiac are apparent:

(1) The astronomical usage of this system purely as a coordinate system, which probably originated with Hipparchus in the second century BC; and

(2) The astrological usage, whereby qualities were attributed to the twelve signs, which were considered to exert an influence occasioned by the passage of the sun, moon, and planets through the zodiacal signs.

As discussed below in relation to the history of the zodiac, the sidereal zodiac was used in astrology throughout the ancient world until—through the influence of Ptolemy—the tropical zodiac was introduced to Arabic astrologers.

Also in modern astronomical tables the ecliptic coordinate system of the tropical zodiac is used as one of the frames of reference for recording the movements of the sun, moon, and planets. For example, *The Astronomical Ephemeris* gives the ecliptic location of the sun each day at midnight in Universal Time (UT) [formerly Greenwich Mean Time (GMT)], expressed in

---

1. Gundel (1978).
2. Neugebauer, *The Exact Sciences in Antiquity*, pp 68–69, indicates the stellar longitudes in this star catalogue to be correct for the time between 130 BC and 60 BC.
3. For example, in the works of the voluminous Arabic astrologer Māshā'allāh (died before AD 830) and also in those of his successor, the influential Arabic astrologer Abū Ma'shar (787–866), all the horoscopes are cast in the framework of the tropical zodiac; cf. Kennedy-Pingree, *The Astrological History of Māshā'allāh* and Pingree, *The Thousands of Abū Ma'shar*.

degrees, minutes, and seconds of arc, from 0° to 360°.[1] Since the vernal point is taken as the zero point on the ecliptic, the longitude of the sun at the moment of the vernal equinox is by definition 0°00'00". In terms of Euctemon's solar calendar, this moment is the start of day 1 of the solar month of Aries. Thirty days later, when the longitude of the sun is 30°00'00", the start of the first day of the solar month of Taurus occurs. Similarly, when the solar longitude is 60°00'00", Gemini 1 starts; and so on (Table 4).

Table 4
Specification of Solar Months in Relation to
the Longitudes of the Corresponding Signs of the Tropical Zodiac

SOLAR MONTH	LONGITUDES OF CORRESPONDING TROPICAL SIGN
Aries	0°00'00" – 29°59'59"
Taurus	30°00'00" – 59°59'59"
Gemini	60°00'00" – 89°59'59"
Cancer	90°00'00" – 119°59'59"
Leo	120°00'00" – 149°59'59"
Virgo	150°00'00" – 179°59'59"
Libra	180°00'00" – 209°59'59"
Scorpio	210°00'00" – 239°59'59"
Sagittarius	240°00'00" – 269°59'59"
Capricorn	270°00'00" – 299°59'59"
Aquarius	300°00'00" – 329°59'59"
Pisces	330°00'00" – 359°59'59"

THE TROPICAL ZODIAC: A UNIVERSAL SOLAR CALENDAR

The solar months derived from Euctemon's tropical calendar comprise a twelvefold division of the year, defined astronomically by the occurrence of the equinoxes and solstices, in which the days were simply counted off (Figure 2). Transforming from the tropical calendar to the tropical zodiac by way of spatial projection (Figures 2 and 10) and referring to 30° arcs along the ecliptic instead of simply counting off the days of the solar months, a schematic solar calendar is obtained (Table 4). In this schematic solar calendar, each month is thirty days long. The days are not actual but ideal, defined astronomically by the length of time taken for the sun's longitude to increase by exactly 1°.

1. *The Astronomical Ephemeris* (yearly).

For example, the first day of Aries is the time interval during which the sun's longitude increases from 0°00'00" to 0°59'59"; and Aries 2 is defined by the increment of solar longitude from 1°00'00" to 1°59'59"; and so on.[1] As well as the Greek astronomer Euctemon, the Babylonians also used a schematic solar calendar, which can be regarded as the original forerunner of the tropical zodiac, as discussed below.

The astronomically defined solar year with twelve months each comprising thirty days indicated in Table 4 is a schematic version of Euctemon's solar calendar (Figure 2). The astronomical division of the year into twelve solar months, associated with the name of Euctemon, can thus be conceived of—in this schematic form—as a universal calendar scheme. In this universal calendar the year has 360 'days', where each day corresponds to a degree of the tropical zodiac and where each 30° arc is named after one of the twelve signs of the zodiac, commencing with Aries for the first 30°, Taurus for the second 30°, etc. This division defines a universal calendar in which each month is measured by the passage of the sun through the corresponding 30° arc in space, and each of the thirty 'days' of the month is measured by the passage of the sun through 1°. This calendar is linked to the astronomically defined occurrences of the solstices and equinoxes, and also to the passage of the seasons, e.g. in the Northern Hemisphere the end of the month of Gemini and the beginning of the month of Cancer coincide with midsummer.

The famous observation made by Euctemon and Meton of the summer solstice in Athens in the year 432 BC serves as an example to illustrate this. The time of the summer solstice in that year was around 10:30 a.m. (local time in Athens) on 28th June 432 BC[2] In terms of this universal calendar, 10:30 a.m. signified the end of day 30 in the solar month of Gemini and the start of day 1 in the solar month of Cancer. Since by definition the summer solstice indicates the start of day 1 in the solar month of Cancer, the longitude of the sun at the start of Cancer 1 was 0° Cancer in the tropical zodiac. It has to be borne

---

1. This concept of 'days' of the solar months is analogous to that of 'tithis', days of the lunar month used in Babylonian and Indian astronomy. A tithi is approximately 24 hours in duration and is defined astronomically by the length of time taken for the elongation of the moon (from the sun) to increase by 12°. There are thirty tithis in a lunar month; the first tithi lasts from new moon (conjunction) until the moon's elongation is 12°. Cf. Pillai, *An Indian Ephemeris*, p 13.

2. Ptolemy, *Almagest* III, 1 (trsl. Toomer, p 138) says that the summer solstice in this year was observed by Meton and Euctemon but only superficially recorded for the morning of Phamenoth 21 of the year 316 in the Era Nabonassar. According to Neugebauer, HAMA II, p 622, this date equated historically with 27th June 432 BC, indicating that Meton and Euctemon were one day in error with their observation of the summer solstice. Note that this date for the summer solstice is given in terms of the Julian calendar that historians use for dates in antiquity, differing from the modern Gregorian calendar in which the summer solstice occurs on (or around) 22nd June.

in mind, however, that the use of the ecliptic coordinate system of the tropical zodiac was at that time not yet known, and so Meton recorded the longitude of the sun to be 8° Cancer, since evidently he was using the sidereal zodiac operating with the norm of System B (vernal point at 8° Aries). On 28th June in that year, which was the *actual* day of the summer solstice, the sun was at 9° Cancer in the reconstructed Babylonian sidereal zodiac (Appendix I), since the vernal point was at 9° Aries in 432 BC. However, owing to a one day error in their determination of the day of the solstice, even though he determined the sidereal longitude of the solstitial point incorrectly (1° error), Meton recorded the longitude of the sun correctly, since the sun was actually at 8° Cancer one day earlier, on 27th June 432 BC.[1] This fact seems to confirm Columella's words concerning Meton and Euctemon as authentic. Not only are they important for an understanding of the history of the tropical zodiac but also indicate that Meton was making observations in the framework of the Babylonian sidereal zodiac as early as 432 BC, providing evidence that the Babylonian sidereal zodiac existed and had been transmitted to Greece prior to this date. Further, since Euctemon and Meton are named together, it is evident that their astronomical school used two systems simultaneously: the tropical calendar of Euctemon and the sidereal zodiac of the Babylonians (System B). The latter was used to specify the location of the summer solstitial point (and, implicitly, the location of the vernal point) in the sidereal zodiac, whereas the former defined a calendar of the year comprising twelve solar months named after the constellations of the zodiac.

THE HISTORY OF THE TROPICAL ZODIAC

Applying the historical approach to help identify the origin of the tropical zodiac, as discussed in Chapter 2 a possible line of transmission was from Euctemon (tropical calendar) via Aratus to Hipparchus (tropical zodiac). Clearly the tropical zodiac must have originated after the time of Meton and Euctemon (fifth century BC), since—if this line of transmission is correct—its innovation depended upon the existence of the division of the year into twelve solar months, which appears to have been introduced by Euctemon. However, the alternative possibility was also discussed in Chapter 2 that the origin of the tropical zodiac with Hipparchus occurred by way of *permanently fixing* the vernal point at 0° of the Babylonian sidereal zodiac at that time on account of his apparent conviction that the vernal point really was at 0° Aries in his day.

Obviously a prerequisite for defining the tropical zodiac was observation of

---

1. Ibid.

the equinoxes and solstices in order to be able to determine the locations of the vernal point, summer solstitial point, etc. Although it was only much later (in Systems A and B of Babylonian mathematical astronomy developed from the fifth century BC onwards, discussed in Chapter 2) that the Babylonians were concerned with the problem of determining the location of the vernal point, etc. in the sidereal zodiac, mention of a scheme for the lengths of daylight and nighttime measured in minas of water at the time of the equinoxes and solstices was made on the Old Babylonian tablet BM 17175 + 17284 stemming from the second millennium BC,[1] indicating that already at this early time the Babylonians were interested in these astronomical phenomena relating to the sun's yearly cycle. And in the Greek tradition of *parapegmata* ('star calendars'), already in Hesiod's poem *Works and Days* (seventh century BC) references to the solstices and equinoxes are made. These references of a calendrical nature, e.g. indicating that the evening rising of Arcturus comes sixty days after the winter solstice,[2] do not imply that at that time there was an accurate astronomical knowledge of the occurrence of the solstices and equinoxes, and probably the equinoxes were known only roughly as the times of the year when the days and nights were approximately equal. Evidently it was not until Hipparchus (second century BC) that the vernal equinox—and correspondingly the vernal point—was able to be specified with sufficient accuracy to warrant choosing it as the primary point of reference for astronomical observations in the form of the tropical zodiac.

One of the most outstanding Greek astronomers prior to Hipparchus was Eudoxus (ca. 408–355 BC), a contemporary of Plato. He is said by Diogenes Laertius to have studied under Plato and later stayed with Egyptian priests at Heliopolis from whom he gained his knowledge of astronomy.[3] Like Euctemon, Eudoxus is famous for his parapegma, or astronomical (star) calendar.[4] However, there is specific evidence from Hipparchus that Eudoxus did not use the tropical zodiac but, like the Babylonians, he used a sidereal zodiac in which the twelve signs, each occupying 30° of celestial space, were marked out by fixed stars. According to Hipparchus, Eudoxus "placed the midpoints (15°) of the signs at these points,"[5] i.e. locating the vernal point at 15° Aries, which is a different norm from that of 8° Aries referred to by Meton and used in the

---

1. Hunger-Pingree (1989), pp 163–164.
2. Hesiod, *Works and Days*, 564–569.
3. Diogenes Laertius, *Lives* VIII, 86–91 (trsl. Hicks II, pp 401–407).
4. Neugebauer, HAMA II, p 588. Eudoxus is also famous as probably the first Greek astronomer to apply a cinematic model to explain the movements of the planets, cf. Neugebauer, HAMA II, pp 675–685. He explained the apparent planetary motions as the consequence of interactions between concentric rotating spheres.
5. Neugebauer, HAMA I, p 278.

Babylonian System B.[1] However, there is an interesting correspondence between this norm utilized by Eudoxus and the schematic solar calendar described in the Babylonian text MUL.APIN. In contrast to the schematic solar calendar of Euctemon described above, in which the solstices and equinoxes occur at the start of the solar months, in the schematic solar calendar of MUL.APIN the equinoxes occur on the 15th days of the first and the seventh months and the solstices occur on the 15th days of the fourth and the tenth months of the year.[2] In their discussion of the schematic solar calendar of MUL.APIN, Hunger and Pingree comment:

> This positioning of the colures on the 15th continued into early Greek astronomy, when Eudoxus, using zodiacal signs instead of months, set the equinoxes in the middle (τὰ μέσα) of Aries and of Libra, the summer solstice in the middle of Cancer, and the winter solstice in the middle of Capricorn.[3]

Hunger and Pingree are probably correct in their conjecture of a direct continuation from the schematic solar calendar of MUL.APIN to the zodiac of Eudoxus with the equinoxes and solstices at 15° of the signs Aries/Libra and Cancer/Capricorn. For exactly this same line of development—but now with the equinoxes and solstices at 0° of the signs Aries/Libra and Cancer/Capricorn—can be seen from Euctemon's solar calendar to the tropical zodiac of Hipparchus.

The schematic solar calendar of MUL.APIN gives us a further clue concerning the origin of the tropical zodiac. It is the statement that the sun stays in three different paths during the cycle of the year:

> the path of Anu during the months XII to II;
> the path of Enlil during the months III to V;
> the path of Anu during the months VI to VIII;
> and the path of Ea during the months IX to XI.[4]

The idealized solar calendar of MUL.APIN can be represented in circular form (Figure 11). What Figure 11 represents astronomically is the movement of

---

1. Neugebauer, ACT I, p72.
2. Hunger-Pingree (1999), p61. Moreover, Pingree notes that there existed an earlier scheme in *Enūma Anu Enlil*, tablets of celestial omens from the first half of the first millennium BC, in which the equinoxes were placed on the 15th days of the twelfth and sixth months and the solstices on the 15th days of the third and sixth months (p50).
3. Ibid., p66. The colures are defined by two great circles passing through the north and south celestial poles, one circle passing through the vernal and autumnal equinoctial points and the other through the summer and winter solstitial points.
4. Ibid., p61. Anu, the sky god, was the supreme god in Mesopotamia until he was supplanted by Enlil ("lord wind"), who was born of the union of Anu ("heaven") and Ki ("earth"). Enlil separated heaven and earth and carried off the earth as his portion. Holding the tablets of

THE TROPICAL ZODIAC   67

the sun in declination during the course of the year. It is the declinational motion of the sun that underlies the concept of the three paths each divided in the way shown in Figure 11. The sun is in the path of Enlil when its declination exceeds 16⅓° north and it is in the path of Ea when its declination is less than 16⅓° south. Between 16⅓° north and 16⅓° south it is in the path of Anu.[1]

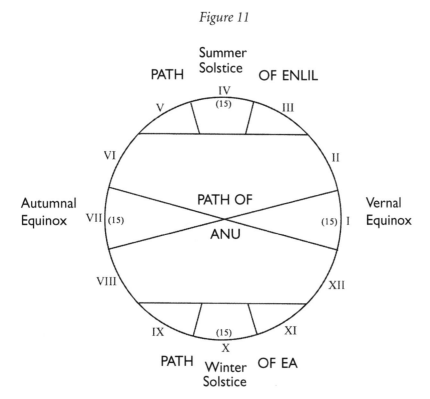

Figure 11

destiny giving him power over human affairs, Enlil was the god of air, wind and storms. Together with Enlil and Anu, Ea, the third in this trinity of ancient gods ruling the Sumero-Babylonian pantheon, was the god of the waters on which the earth floats. He was a god of wisdom, magic, arts and crafts. The mythological introduction to the omen tablets *Enūma Anu Enlil* referred to in n16 traces the order of the heavens and of the earth back to the gods Anu, Enlil, and Ea; cf. Koch-Westerholz (1995), pp77–78.

1. 16⅓° is the precise value. Hunger and Pingree state it to be ca. 15°: "The Path of Anu seems to occupy the arc of the horizon over which stars with declinations between ca. 15° N and 15° S rise, the Path of Enlil the arc on the horizon north of this, the Path of Ea the arc south of this" (ibid., p61).

68   HISTORY OF THE ZODIAC

To make this more explicit, Figure 12 indicates the values in declination for the sun's passage through each month. For example, the sun in this schematic calendar would have risen from 6° south to 6° north of the celestial equator during the first month of the year, the month of Nisannu, which corresponded to the Babylonian constellation of the Hired Man, in turn corresponding to the constellation of Aries the Ram.[1] And according to the following text: "Nisan is the month of the constellation Ikû (Aries), which is the throne-room of Anu. The king is lifted up, the king is installed. The blessed springing forth of vegetation of (by) Anu and Enlil."[2]

Figure 12

The Sun's Motion in Declination in Relation to the Schematic Solar Months of the MUL.APIN Calendar

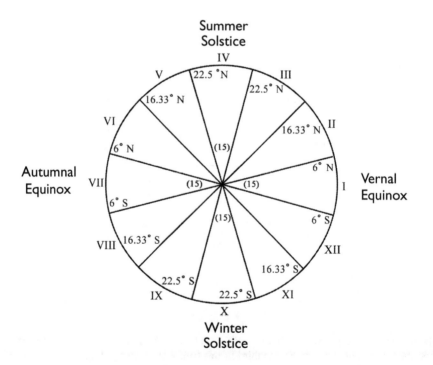

---

1. Hunger and Pingree indicate that the Babylonians, according to one of the star lists belonging to MUL.APIN, observed the heliacal rising of the Hired Man during the first 20 days of Nisannu (ibid., p 65). They identify the Hired Man with Aries (ibid., p 273).
2. Langdon, *Babylonian Menologies and the Semitic Calendars*, p 68.

As a second example from Figure 12: during the month of Arahsamnu (month VIII) the sun in this schematic solar calendar would have descended from 6° to 16⅓° south. Evidently the month Arahsamnu corresponded to the constellation of the Scorpion.[1]

It emerges from MUL.APIN, based on the heliacal risings of constellations during the cycle of the twelve months of the year, that a correspondence was established between the constellations and the twelve months of the year. Of course, the primary calendar of Babylonian civilization was a lunar calendar, but the same principle of a correspondence of the twelve signs/constellations of the zodiac with the twelve months of the schematic solar calendar of MUL.APIN seems to hold true also for the lunar calendar, extrapolating from the correspondence Nisan—Aries (above quotation) to the rest of the lunar months of the year. Evidently this correspondence arose originally through observing the heliacal rising of the constellations during the cycle of the twelve months of the year.

From approximately −500 onwards the Babylonians used the 19-year lunar cycle to regulate their calendar.[2] In this calendar the vernal equinox generally fell during month XII (Addaru), although sometimes it fell near the beginning of month I (Nisannu).[3] Since the Babylonian months began with the *visible new moon*, i.e. the first appearance of the new crescent moon on the western horizon after sunset, which could occur up to two days after the *astronomical new moon* (conjunction of sun and moon), it is clear that the new moon falling prior to the vernal equinox specified the start of the twelfth month (Addaru), although sometimes it was the new moon falling immediately after the vernal equinox. This meant that the new moon at the start of the first month (Nisannu) should always fall after the vernal equinox. Since the vernal point was located at 10° Aries in −500 (Appendix I), then the month of Nisannu would generally begin when the sun was well advanced in the sidereal sign of Aries. Since the distance between the sun and a fixed star was reckoned to be around 15° in order for a first magnitude star to be visible prior to sunrise (heliacal rising, see Figure 15),[4] and the interval had to be correspondingly greater for less bright stars, clearly the sun had to be located at least 15° into the

---

1. Hunger and Pingree indicate that the Babylonians observed the heliacal rising of the Scorpion during most of the month of Arahsamnu, from the fifth day onwards (Hunger-Pingree [1999], p65).
2. Neugebauer, HAMA I, p355 indicates that the 19-year intercalation cycle extends almost continuously from −497 onwards.
3. Ibid., p362.
4. Neugebauer, HAMA I, p261, indicates that 15° was considered by Ptolemy to be the normal 'arcus visionis' for first magnitude stars.

sidereal sign of Aries in order for the stars at the beginning of the sign of Aries to be visible before dawn, and since the brightest star in the first eight degrees of Aries is Alpherg (η Piscium)[1], which is actually rather faint (magnitude 3.62), probably the sun had to be located at 19° Aries, which was the case some nine days after the vernal equinox. Normally this condition was fulfilled in the month of Nisannu of the Babylonian calendar, so that during this month the heliacal rising of the Hired Man (Aries the Ram) was observed, which was the basis for the correspondence between Nisannu and Aries.

To summarize: in the post MUL.APIN astronomy, once the twelve signs/constellations of the Babylonian sidereal zodiac were identified, through observation of the heliacal rising of these signs/constellations during the lunar months, the following correspondences were established: the first month of the year (Nisannu) was seen to correspond to the Hired Man (Aries), the second month (Ajjaru) to the Bull of Heaven (Taurus), the third month (Simānu) to the Great Twins (Gemini), the fourth month (Du'ūzu) to the Crab (Cancer), the fifth month (Abu) to the Lion (Leo), the sixth month (Ulūlu) to the Furrow (Virgo), the seventh month (Tešrītu) to the Scales (Libra), the eighth month (Arahsamnu) to the Scorpion (Scorpio), the ninth month (Kislīmu) to the constellation of Pabilsag (Sagittarius), the tenth month (Tebētu) to the Goat-Fish (Capricorn), the eleventh month (Šabātu) to the constellation of the Great One (Aquarius), and the twelfth month (Addaru) to the constellation of Anunītu (Pisces).[2]

These correspondences between months and signs/constellations of the zodiac were based on observation of the heliacal risings of the signs/constellations of the zodiac. Such correspondences stating in which month a certain constellation rises, although generally astronomically incorrect, are of great antiquity and are attested already in the series of tablets *Enūma Anu Enlil* of celestial omens for the end of the second millennium BC and the first half of the first millennium BC.[3] The practice of equating months and signs was continued in MUL.APIN and in other texts from the Neo-Babylonian period. "The system used for grouping the months was transferred to zodiacal signs. Such equivalence of months and signs can be found in several Babylonian

---

1. Appendix I indicates that the longitude of Alpherg in the sidereal zodiac is 2° Aries. Alpherg is usually included with the constellation of Pisces, but Hipparchus referred to it as "the star marking the forefoot of the Ram" (*Commentary to Aratus,* trsl. K. Manitius, p255).

2. Hunger-Pingree (1999), pp271–277, lists the Babylonian stellar constellations in relation to modern astronomical constellations (but does not indicate a correspondence with the twelve months of the Babylonian calendar).

3. Reiner-Pingree (1981), Tablet 51, Section 1. Tablet 51 relates to stars listed in the Old Babylonian astrolabes going back to about –1100.

texts from the Neo-Babylonian period."[1] As discussed above, these observations (and the resulting correspondence with the months) depended on the occurrence of the vernal equinox during month XII (Addaru)—or, exceptionally, close to the beginning of month I (Nisannu). Here a difference is apparent in comparison with the schematic solar calendar of MUL.APIN in which the vernal equinox was located in the middle of month I (Nisannu). During the MUL.APIN period of Babylonian astronomy—the oldest copy being dated to –686, although containing observations based on substantially older sources—the twelve signs/constellations of the sidereal zodiac had not yet been clearly identified, and so there was no question of establishing a correspondence with the months such as took place later, after the twelve signs/constellations were defined. Based on the reconstruction of the Babylonian sidereal zodiac given in Appendix I, the vernal point in the middle of the seventh century BC, during the MUL.APIN period, was approximately 12° Aries, since the time interval from –650 to AD 220 was 870 years (12 x 72 = 864). This position of the vernal point close to the center of sidereal Aries theoretically would explain why the first month (Nisannu) of the idealized solar calendar was defined with the vernal point at its center (on the 15th day of the idealized month of thirty days), but a practical observational foundation for this is lacking, signifying that—at the present point of time (lacking any textual evidence)—it is not possible to really explain why the Babylonians placed the vernal equinox at the middle of the month of Nisannu in their schematic solar calendar. According to the reconstructed Babylonian sidereal zodiac in Appendix I, the vernal point was located at 15° Aries in –860, occurring 1080 years before AD 220 (15 x 72 = 1080). It was conceivably around –860 that MUL.APIN—and its schematic solar calendar—originated.

Eudoxus' adoption of the sidereal zodiac with the vernal point at 15° Aries was perhaps directly connected with knowledge of this schematic solar calendar in MUL.APIN, as Hunger and Pingree point out.[2] In this case Eudoxus possibly heard from sources familiar with Babylonian astronomy about their zodiac with twelve equal (30°) signs and about their solar calendar in which the vernal equinox was located in the middle of the first month, and also that the first month corresponded to the first zodiacal sign (Aries). In this case an obvious step, then, would be to conclude that the vernal point should be located at 15° Aries. And not knowing about precession, he probably believed that the vernal point was fixed at 15° Aries.

Moreover, it was evidently the principle of a correspondence between the twelve months of the yearly cycle and the twelve zodiacal constellations which

1. Hunger-Pingree (1999), p17.
2. Hunger-Pingree (1999), p66.

underlay the association of the solar months of Euctemon's calendar with the twelve signs/constellations of the zodiac. Returning to Columella's statement concerning Meton and Euctemon, the two astronomers evidently each represented a different astronomical principle, perhaps in each instance adopted from Babylonian astronomy. In the case of Meton, who according to Columella observed the summer solstice in 432 BC to be at 8° Cancer (and thus the vernal point at 8° Aries) it was the Babylonian zodiac with the norm of System B (vernal point at 8° Aries) which he had assimilated.[1] If Meton really had assimilated the Babylonian sidereal zodiac (rather than making an independent discovery of a sidereal zodiac with the vernal point at 8° Aries), he must have had access to Babylonian sources. This conjecture is strengthened by the fact that Meton utilized a 19-year lunar calendar almost identical to the one that the Babylonians had been using since around 490 BC.[2] The calendar cycle attributed to Meton came to be known as the *Metonic cycle*.

With regard to Meton's associate, it can be conjectured that just as Meton had adopted the 19-year Babylonian lunar calendar, Euctemon had assimilated the schematic solar calendar of the Babylonians and knew of the correspondence between the twelve months and the twelve signs/constellations of the zodiac. In this case, then, why did he devise a calendar scheme with the months commencing with the equinoxes and solstices, so that the month of Aries began at the vernal equinox instead of the vernal equinox occurring on the 15th day as in the Babylonian solar calendar? It was, of course, still the sun's motion in declination underlying his calendar, but with each month shifted in relation to the original solar calendar of the Babylonians from MUL.APIN. Thus Euctemon's tropical calendar as the predecessor of the tropical zodiac is clearly nothing other than a model of the sun's yearly movement in declination, with the sign or month of Aries defined as the period of time when the sun's declination increases from 0° to 11½°N, the sign or month of Taurus when the sun's declination increases from 11½°N to 20°N, etc. The sun's movement in declination in the framework of Euctemon's solar calendar is shown in Figure 13.

---

1. Either he knew about the Babylonian sidereal zodiac and utilized it in his astronomical observation or, if he did not know about it, he discovered the principle of the Babylonian zodiac (with the vernal point at 8° Aries) for himself. In both cases this means that he had assimilated the principle of the sidereal zodiac as an ecliptic coordinate system, but only in the former instance is it possible to speak of a Babylonian influence upon the definition of the zodiac during this early phase (fifth century BC) of the development of Greek astronomy.

2. Neugebauer, HAMA II, pp 622–623, points out that there are some differences between Meton's version of the 19-year lunar calendar cycle and the Babylonian original.

Figure 13

Sun's Motion in Declination
in Relation to the Solar Months of Euctemon's Tropical Calendar

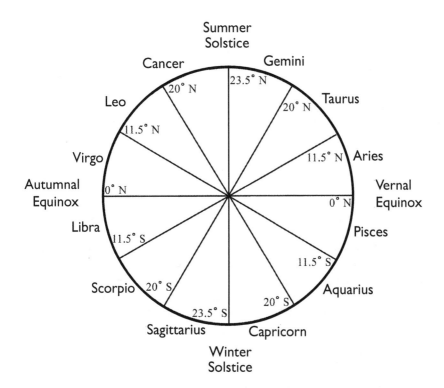

If Eudoxus somehow knew of the schematic solar calendar of the Babylonians, it is not too far-fetched to postulate that Euctemon came to hear of it too and was thus inspired to introduce a solar calendar to the Greeks. But whereas Eudoxus held to the specification of the vernal equinox occurring in the middle of Nisannu (the month of Aries), which he translated into a projection of the vernal point to the middle (15°) of the corresponding sign/constellation of the Ram, Euctemon opted to re-define the solar calendar with the vernal equinox occurring at the beginning of the month of Aries. This might have been purely for practical reasons. To begin counting days at the vernal equinox was simpler than trying to count in such a way that the vernal equinox took place on the 15th day of the month. The fact that he named the months thus obtained (by counting days from the solstices and equinoxes) after the signs of the zodiac shows that he was aware of the correspondence between the months and the zodiacal signs.[1] And it was this, evidently, that proved decisive for Hipparchus in his adoption of the tropical zodiac (considered as a spatial projection of Euctemon's tropical calendar, just as Eudoxus apparently specified a spatial projection of the schematic solar calendar from MUL.APIN). In the case of Eudoxus' spatial projection of the schematic solar calendar of the Babylonians, the vernal point was placed in the middle of the sign of Aries (15° Aries), and in the case of Hipparchus' spatial projection of Euctemon's solar calendar the vernal point was located at 0° Aries. From the reconstructed Babylonian sidereal zodiac (Appendix I) this location of the vernal point did not take place until AD 220, around three and one-half centuries after Hipparchus. Similarly, the placing of the vernal point at 15° Aries by Eudoxus would have been an accurate position five centuries before him, in −860, as mentioned above.

FURTHER HISTORICAL REMARKS

Considering the importance of Columella's statement concerning Meton and Euctemon in all these considerations, here it is quoted more fully:

> From the setting of the Pleiades till the winter solstice, which falls about December 23rd in the eighth degree of Capricorn.... I am well acquainted with the reckoning of Hipparchus, which declares that the solstices and equinoxes occur not in the eighth but in the first degrees of the signs of the Zodiac; however, in these rural instructions I am now following the calendar of Eudoxus and Meto and the old astronomers, which are adapted to the public festivals, because this view, accepted in old times, is more familiar to farmers and, on the other hand, the subtlety of Hipparchus is not necessary for rustics of less refined education.[2]

Columella, writing in the first century AD, provides literary evidence that the early Greek astronomers Meton and Eudoxus used the sidereal zodiac of the Babylonians. From Columella's statement it can also be inferred that Hipparchus (second century BC) introduced a reform of the zodiac, which was still a novelty in the first century AD. According to this testimony, Hipparchus' zodiacal reform was to define the zodiac with respect to the tropical (solstitial) and equinoctial points. This redefinition of the zodiac meant that the fixed stars were no longer used as an astronomical coordinate system for locating the positions of the sun, moon, and planets. Whereas previous astronomers—notably the Babylonians—divided the fixed star zodiac into twelve 30° signs

---

1. This correspondence was evidently arrived at by the Babylonians by way of observing the heliacal rising of the zodiacal signs/constellations during the months of the Babylonian lunar calendar; cf. Hunger-Pingree (1999), p 65.
2. Columella, *De re rustica* IX, 14 (trsl. Ash-Forster-Heffner II, pp 487–489).

and identified planetary positions by measuring the distances of the planets from the fixed stars marking the sidereal signs, Hipparchus adopted the practice of measuring the distances of planets from the tropical (solstitial) and equinoctial points, evidently for practical reasons. The previous astronomical practice became completely reversed. In the sidereal coordinate system of the Babylonians—adhered to also by the Greek astronomers Meton and Eudoxus (according to Columella)—not only the planets were located by referring to their positions in the twelve fixed star signs, but also the solstices and equinoxes were placed in relation to the degrees of the sidereal signs. Hipparchus abandoned the coordinate system of the sidereal signs altogether and instead began to make measurements in relation to the tropical and equinoctial points. According to van der Waerden the development from the sidereal to the tropical zodiac was as follows:

> The Babylonian division of the zodiac is *sidereal*, i.e. the limiting points of the signs have a fixed position with respect to the fixed stars. For instance, Spica always lies near the end of the sign of Virgo. In this system, the longitudes of the fixed stars are constant: they were given in a 'star catalogue', which was used in ancient and in more recent times. Longitudes of the moon and the planets were determined by observing their distances to fixed stars. Longitudes of the sun were probably determined by observing eclipses. On the other hand, most Greek authors define the zodiacal signs by means of the equinoxes and solstices. The initial points of Aries, Cancer, Libra and Capricorn are, by definition, the points at which the sun's center stands at the equinoxes and solstices. This is the *tropical division* of the zodiac. A consequence of these definitions is a quite different order of the observations and a different structure of the theory. Hipparchus started with accurate determinations of equinoxes and solstices. Next he determined the eccentric motion of the sun.... From the longitude of the sun, the longitude of the moon can be found, e.g. by observing eclipses. Longitudes of planets and fixed stars are determined by observing their distances from the moon.[1]

The fact that the tropical coordinate system introduced by Hipparchus was adopted by Ptolemy ensured that it became the primary reference system in astronomy from the time of Ptolemy (second century AD) onwards. Consequently the old sidereal division into twelve equal fixed-star signs fell into disuse both observationally and computationally. The sidereal zodiac was *simultaneously* the frame of reference for the observations of the Babylonian astronomers and also their coordinate system for computing the locations of the sun, moon, and planets in the signs of the zodiac. The observational function of the Babylonian sidereal zodiac was replaced by the astronomical zodiac

---

1. Van der Waerden, SA II, pp 288–289.

divided into twelve unequal constellations, in which the constellations retained the same names as the sidereal signs. And the computational function of the sidereal zodiac was replaced by the tropical zodiac in which the signs received the same names as the signs of the sidereal zodiac. The transmission of the Babylonian sidereal zodiac to Egypt and India is examined in Chapters 4 and 6.

When Ptolemy drew up his famous catalogue of stars, he expressed the longitudes of fixed stars in terms of the signs of the tropical zodiac. For example, the star λ Leonis which Ptolemy refers to as the one "in the gaping jaws" of the Lion he measured to be a distance of 21⅙° eastward of the summer solstice point.[1] Since the tropical sign of Cancer is defined to be the arc of celestial space extending 30° east of the summer solstice point, he recorded the longitude of λ Leonis to be 21⅙° Cancer. In this instance a star which in the Babylonian sidereal coordinate system belonged *by definition* to the sign of Cancer became catalogued by Ptolemy as belonging to the constellation of Leo. It might be concluded that the star configuration called Leo in the Babylonian sidereal zodiac, occupying 30° of celestial space, thereby became extended in length by Ptolemy to incorporate some of the stars that previously belonged to Cancer—thus correspondingly diminishing Cancer in size—and that the same took place at other constellation boundaries, with stars becoming redefined in terms of belonging to a particular sign/constellation of the zodiac.

However, it is evident that prior to the introduction of the ecliptic coordinate system of the sidereal zodiac, Babylonian astronomers regarded the constellations as being unequal in length. Thus, for them, the star Rasalasad (ε Leonis) was initially referred to as the Head of the Lion in the Normal Star system.[2] Later, after the introduction of the ecliptic coordinate system of the Babylonian sidereal zodiac, it probably received the longitude 26° Cancer (Appendix I). Similarly, the star Zavijava (β Virginis) was referred to in the Normal Star system as the Rear Foot of the Lion.[3] When this star was assigned a longitude in the Babylonian sidereal zodiac, it was probably 2° Virgo (Appendix I).

On the other hand let us consider the star Alpherg (η Piscium). In terms of the system of Normal Stars that preceded the Babylonian sidereal zodiac, Alpherg was referred to as the Bright Star of the Fishes' Ribbon.[4] In the reconstructed Babylonian zodiac (Appendix I), Alpherg's sidereal longitude is 2° Aries and thus for the Babylonians this star evidently then belonged to the

---

1. Ptolemy, *Almagest* VII, 5 (trsl. Toomer, p367).
2. Hunger-Pingree (1999), p148.
3. Ibid., p149.
4. Ibid., p148.

sign/constellation of Aries. For Ptolemy, over five hundred years later, Alpherg belonged to Pisces. However, here Ptolemy differed from his forerunner Hipparchus who, in line with the later Babylonian specification in the sidereal zodiac, called this "the star marking the forefoot of the Ram."[1] In consideration of this star Alpherg, it can be conjectured that the (no longer extant) star catalogue of Hipparchus was perhaps closer to the Babylonian sidereal zodiac than Ptolemy's catalogue in the *Almagest*, since it is known that Hipparchus adopted Babylonian astronomical parameters.[2] Moreover, Ptolemy explicitly states that the "descriptions which we have applied to the individual stars as parts of the constellation are not in every case the same as those of our predecessors."[3] From this statement by Ptolemy, since the only predecessor whom he explicitly names is Hipparchus, it would seem that Ptolemy's catalogue of stars represents a re-formation of that of Hipparchus (and, ultimately, of that of the Babylonians) to yield the twelve unequal-length astronomical constellations in place of the equal-length constellations/signs of the Babylonian sidereal zodiac. The first stage of this re-formation might have been the (lost) catalogue of Hipparchus in which the stellar locations perhaps differed only slightly from the original Babylonian star catalogue,[4] and that this difference increased, i.e. a more substantial re-formation took place, when Ptolemy compiled his star catalogue.

Comparing the modern astronomical zodiac (based on Ptolemy's star catalogue, as discussed in Chapter 1) with the Babylonian sidereal zodiac: with the exception of the boundary between Virgo and Libra, there is a reasonable agreement in the stellar locations of the boundaries between the sidereal signs and the constellation figures of Ptolemy (Figure 14).

The foregoing discussion relating to the comparison of Ptolemy's star catalogue and the lost Hipparchian star catalogue with the star catalogue defining the Babylonian sidereal zodiac applies only to the twelve zodiacal constellations comprising the astronomical zodiac, since there is no evidence that the star catalogue defining the Babylonian sidereal zodiac contained any stars beyond the immediate vicinity of the ecliptic. In fact, the latitudes of the 32 Normal Stars which then became assigned longitudes in the Babylonian sidereal zodiac all lie between 10° north and 7½° south of the ecliptic, i.e. within the zodiacal belt.[5]

1. Hipparchus, *Commentary to Aratus* (trsl. K. Manitius, p 255).
2. Toomer (1988).
3. Ptolemy, *Almagest* vii, 4 (trsl. Toomer, p 340).
4. Sachs (1952, 2).
5. Ibid., pp 148–149.

78 HISTORY OF THE ZODIAC

The system of twelve unequal fixed-star constellations (astronomical zodiac) replaced the twelve equal 30° signs (Babylonian sidereal zodiac). What was for the Babylonians *one* system, a sidereal coordinate system used for both observing and measuring planetary positions, became through Hipparchus and Ptolemy *two* systems, a *tropical coordinate system* for measuring planetary positions and a *sidereal division* into twelve unequal constellations used for observation. It is these two systems of Hipparchus and Ptolemy that modern astronomy has inherited and uses today.

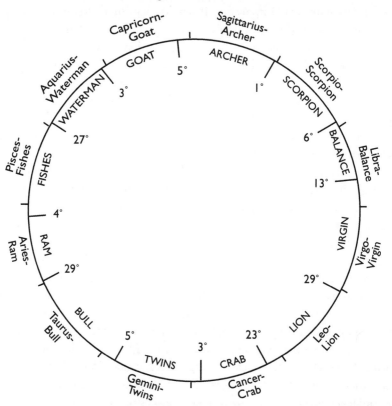

Figure 14
Sidereal Signs and Constellations

Figure 14 compares the equal-division Babylonian zodiac (fifth century BC) with the modern unequal division zodiac with constellation limits as adopted by the International Astronomical Union.[1] The figures give the locations (rounded to whole degrees) of the modern zodiacal constellations in relation to the twelve fixed signs of the sidereal zodiac, e.g. the modern Aries (Ram) extends from 4° to 29° of the Babylonian Aries-Ram.

1. Delporte, *Délimitation scientifique des constellations*.

As discussed in Chapter 1, historical analysis shows that Ptolemy may be regarded as the originator of the astronomical division into twelve unequal zodiacal constellations in the sense that the modern scientific delimitations of these constellations—at least of the 48 classical constellations (including the twelve constellations of the astronomical zodiac)—are ultimately based on Ptolemy's catalogue. Similarly Hipparchus appears to be the innovator of the tropical zodiac as a coordinate system since he was, as far as can be ascertained, the first astronomer to make astronomical measurements in the ecliptic coordinate system of the tropical zodiac. Following Hipparchus, Ptolemy used the tropical coordinate system in his astronomical work summarized in the *Almagest*. Ptolemy was also responsible for the introduction of the tropical zodiac into astrology through the *Tetrabiblos*.

THE TROPICAL ZODIAC IN ASTROLOGY

If we look at the astrological textbook of Claudius Ptolemy, the *Tetrabiblos*, written in Alexandria around AD 150, it becomes apparent that he was a gifted astronomer interested in astrology more from an academic standpoint than as a practicing astrologer. Thus the *Tetrabiblos* contains no horoscopes, and there is no evidence that Ptolemy ever cast a horoscope. The practice of casting horoscopes had spread from Babylon to Hellenistic Egypt, Greece, and Rome during the preceding centuries, and was transmitted to India at about the time of Ptolemy. At that time all horoscopes were cast "Babylonian style" in terms of the sidereal zodiac. However, in the course of time knowledge of the definition of the sidereal zodiac was lost. It would seem that Ptolemy did not know exactly how the twelve 30° signs of the zodiac were defined in relation to the fixed stars, but nevertheless, it is clear from the following passages that he regarded the zodiacal signs as sidereal:

> The sign of Taurus as a whole is indicative of both temperatures and is somewhat hot; but taken part by part, its leading portion, particularly near the Pleiades, is marked by earthquakes, winds, and mists; its middle moist and cold, and its following portion, near the Hyades, fiery and productive of thunder and lightening.[1]
>
> For blindness in one eye is brought about when the moon ... applies to one of the star clusters in the zodiac, as for example to the cluster in Cancer, and to the Pleiades of Taurus, to the arrow point of Sagittarius, to the sting of Scorpio, to the parts of Leo around the Coma Berenices, or to the pitcher of Aquarius.[2]

---

1. Ptolemy, *Tetrabiblos* II, 11 (trsl. Robbins, pp 201–203).
2. Ibid., III.12 (trsl. Robbins, p 321).

In the reconstructed Babylonian sidereal zodiac the Pleiades are located at 5° Taurus and the primary star in front of the Hyades, Aldebaran, is at 15° Taurus (Appendix I). The above passage by Ptolemy holds true in the framework of the sidereal zodiac but does not make sense in terms of the tropical zodiac.[1] However, it was Ptolemy himself who introduced the tropical zodiac into astrology.[2] In the *Tetrabiblos* he wrote: "Although there is no natural beginning of the zodiac, since it is a circle, they assume that the sign which begins with the vernal equinox, that of Aries, is the starting-point of them all...."[3] At the time he wrote this (ca. AD 150), the vernal point was located at 1° of the sign of Aries in the sidereal zodiac (Appendix I), making it quite accurate (to within a degree) to say that the sign of Aries began *"with the vernal equinox."* But owing to the precession of the equinoxes, the vernal point has now shifted back from 1° Aries to 5° Pisces, signifying that the sign of Aries now no longer begins with the vernal equinox (20th/21st March) but about twenty-five days later (15th April).[4] Ptolemy, writing in AD 150, was actually quite accurate in defining the sign of Aries to begin with the vernal point, considering that at his time the vernal point was at 1° Aries. However, by putting this down in writing in such a way as to give the impression that this was something permanent, he essentially created a new astrological zodiac. What were the steps involved in the creation of this new astrological zodiac?

THE SUBSTITUTION OF THE TROPICAL ZODIAC FOR THE SIDEREAL ZODIAC

As discussed above, the tropical zodiac is essentially a spatial projection of the tropical calendar. At the time of Ptolemy this projection more or less coincided with the constellational division of the sidereal zodiac of the Babylonians. Initially astrologers paid no attention to Ptolemy's new astrological creation of the tropical zodiac.[5] This has been demonstrated in studies of Greek horoscopes spanning the first to the fifth centuries AD, compiled by Neugebauer and van Hoesen in the publication *Greek Horoscopes*. From

---

1. For example, at the present time Aldebaran's tropical longitude is 10° Gemini.
2. North, *The Fontana History of Astronomy and Cosmology*, p 67.
3. Ptolemy, *Tetrabiblos* 1.10 (trsl. Robbins, pp 59–61).
4. From a present day standpoint 25 days, from 21st March to 15th April, is the time it takes for the sun to pass from its current location at the vernal equinox (5° Pisces) to 0° Aries of the sidereal zodiac at a rate of approximately 1° per day for the motion of the sun.
5. Here a distinction is being drawn between the use of the tropical zodiac as an astronomical coordinate system introduced by Hipparchus and the astrological use of the tropical zodiac in which it is considered, over and above its function as a coordinate system, to possess astrological qualities or influences expressed in its twelve signs.

Neugebauer's analysis it is evident that most of the Greek horoscopes were sidereal, using the same (or similar) frame of reference as the Babylonian sidereal zodiac.[1] However, he distinguishes between the horoscopes found in literary sources and other horoscopes ('original documents') mainly written on papyrus, and concludes that the later literary horoscopes, almost all from the fifth century AD, were in terms of the tropical zodiac.[2] The 'original documents' extend over the period from 62 BC to AD 478, and the literary horoscopes from 72 BC to AD 621. The bulk of these approximately 180 Greek horoscopes are distributed over the first to the fifth centuries.

An example of a literary horoscope is one cast for 28th October 497 in which all the planetary longitudes are between 2° and 4° less than those computed by Neugebauer using the tropical zodiac. Even allowing for computational errors on the part of the astrologer who cast the horoscope, it is evident that a sidereal frame of reference was used.[3] The literary horoscope from AD 516 is too incomplete to enable an assessment concerning whether it was cast in the sidereal or tropical zodiac. The last literary horoscope is for the birth of the new religion of Islam, cast for 1st September 621, the beginning of the Byzantine year preceding the Hijra. Here again the values computed by Neugebauer in the tropical zodiac are greater than those given in the horoscope, again pointing to a sidereal frame of reference.[4] This horoscope was cast by the astrologer Stephanus, and his comments attached to the horoscope reveal a knowledge of events up to AD 775, thus indicating that this horoscope for AD 621 was cast over one and a half centuries later than this date.[5]

An independent analysis by Kollerstrom of the computations upon which 19 Greek horoscopes were based (together with 5 Babylonian horoscopes included in the analysis) leads to the conclusion that:

---

1. Neugebauer-van Hoesen, *Greek Horoscopes*, p 172.
2. Ibid., pp 171–172.
3. If this horoscope were cast in the framework of the Babylonian sidereal zodiac, the amount in 497 that sidereal longitudes would be less in value than the tropical longitudes would be 3°51', since this is the amount of precession between 220 and 497.
4. If this horoscope were cast in the framework of the Babylonian sidereal zodiac, the amount in 621 that sidereal longitudes would be less in value than the tropical longitudes would be 5°41', since this is the amount of precession between 220 and 621. However, the planetary longitudes in this horoscope (excluding Venus' and Mercury's longitudes which are grossly inaccurate) are between only 1° and 4° less than the computed value, and in the case of the moon the longitude given on the horoscope is 2° *more* than the computed value. Thus it seems unlikely that this astrologer was using the Babylonian norm for the sidereal zodiac, and perhaps he was using another sidereal norm. However, it is also conceivable that he was using tropical longitudes but made a consistent error in his computations so that they all (apart from the moon) turned out to be less than they should have been.
5. Neugebauer-van Hoesen, *Greek Horoscopes*, p 190.

82 HISTORY OF THE ZODIAC

A single frame of reference for the sidereal zodiac was used by both Babylonian and Greek astrologers, enduring over eight centuries, before being forgotten in the Dark Ages.... These horoscopes show that even in the centuries after Ptolemy, the astrologers writing in Greek continued to use a sidereal reference. The horoscopes are mainly from Alexandria, indicating that even in Ptolemy's native city the sidereal tradition endured.[1]

With the expansion of Christianity in the West, the practice of astrology in the Christian world declined, effectively disappearing by the sixth century AD, except for isolated examples such as the horoscope from AD 621 for the birth of Islam. The last definitely sidereal horoscope from *Greek Horoscopes* is dated to the year AD 497.[2] Soon after this astrology—as the practice of casting horoscopes—was taken up in the Arabic speaking world with the first Arabic horoscope dating to the year 531.[3] The available evidence, as far as one can tell,[4] shows that Arabic astrologers used the tropical zodiac in casting horoscopes.[5]

When Ptolemy's *Tetrabiblos* was translated from Greek, it became a standard reference work on astrology for the Arabic astrologers, who adopted Ptolemy's introduction of the tropical zodiac into astrology uncritically, not knowing that there had ever been an astrology based on the sidereal zodiac prior to Ptolemy. Consequently the sidereal zodiac was forgotten. Evidently its very existence was unknown to the Arabic astrologers.

Thus, as astrology based on the sidereal zodiac disappeared from the Christian world of the West, at the same time a new kind of astrology based on the tropical zodiac and the work of Ptolemy began to grow and flourish in

---

1. Kollestrom (1997), p15; Kollerstrom used 19 horoscopes from Neugebauer-van Hoesen *Greek Horoscopes* and an additional 5 Babylonian horoscopes later published by Rochberg (1998) spanning the years 235 to 69 BC. He analysed the longitudes of planets (excluding Mercury for which the values are less reliable) from each horoscope. Leaving out obviously incorrect values, his analysis is based on 96 longitudes from which he concludes that the sidereal zodiac was the sole frame of reference used by Babylonian and Greek astrologers. Kollerstrom's analysis shows a best fit for a zodiacal framework in which the star Spica is located at 30° Virgo as in the Indian zodiac (chap. 6). Cf. also Kollerstrom (2001).

2. As discussed in n4 on p81, although the likelihood is that the horoscope of AD 621 was sidereal, using a different norm than the Babylonian one, it is also possible that it was an inaccurately cast tropical horoscope, so this example is not absolutely conclusive.

3. Neugebauer-van Hoesen, *Greek Horoscopes*, p161: "The earliest Arabic horoscope, as far as known to us, is cast for the coronation of the Sasanian King Khosro Anōsharwān (18th August AD 531).

4. It is not always easy, as can be seen from the discussion of the horoscope of the birth of Islam (cast for 1st September AD 621) discussed above, to assess whether a horoscope is using tropical coordinates or a sidereal frame of reference.

5. Neugebauer-van Hoesen, *Greek Horoscopes*, p172; cf. also Pingree, *The Thousands of Abū Ma'shar* and Kennedy-Pingree, *The Astrological History of Māshā'allāh*.

the Arabic speaking world. This new astrology was then introduced into Christian Europe from about 1140 onwards through translations from Arabic into Latin.[1] In this way the tropical astrology of the Arabic astrologers became established in Europe in place of the ancient sidereal astrology that had flourished for at least nine hundred years: from 410 BC, the date of the world's oldest horoscopes found so far,[2] to AD 497, the date of one of the last Greek horoscopes that is known definitely to be sidereal, as discussed above. The extraordinary thing is that apparently no one in the Arabic speaking world or in Christian Europe knew that there had been any other kind of astrology than that of Ptolemy based on the tropical zodiac.[3] And it was not until the twentieth century that knowledge of the existence of the Babylonian sidereal zodiac re-emerged in the West through the excavation and decipherment of cuneiform texts.

1. Tester, *A History of Western Astrology*, p147.
2. Rochberg (1998).
3. Evidently Ptolemy did not consciously introduce the tropical zodiac into astrology through the *Tetrabiblos*, since—as discussed above—his own references in this work indicate a sidereal frame of reference. Rather, his introduction of the tropical zodiac into astrology was a historical accident occasioned by the fact that at the time he wrote the *Tetrabiblos* (ca. AD 150) the vernal point was at 1° Aries in the Babylonian sidereal zodiac, and so Ptolemy's placing of the vernal point at 0° Aries was fairly accurate. The historical accident consisted in this placing of the vernal point at 0° Aries becoming later interpreted as something permanent and unchanging.

4

THE SIDEREAL ZODIAC IN EGYPT

THE CELESTIAL SIGNS OF THE ZODIAC, depicted on modern star maps as the twelve star configurations of the Ram, the Bull, the Twins, etc., are more or less the same as those up to which the Egyptians of the Hellenistic and Roman periods gazed. This is apparent from such archeological findings as the circular zodiac found on the ceiling of the Temple of Hathor at Dendera (late Ptolemaic, but prior to 30 BC), now in the Louvre Museum,[1] and also the rectangular zodiac from the Temple of Khnum (ca. 246–180 BC), which unfortunately was completely destroyed shortly after it was found.[2]

The Dendera zodiac represents very clearly the twelve star configurations of the zodiac, showing the planets in their signs of exaltation.[3] This indicates a definite influence of Babylonian astronomy on that of the Egyptians during the Hellenistic period, since the doctrine of exaltations is of Babylonian origin going back to the seventh century BC.[4] There are minor differences in some of the figures in the Dendera zodiac from those described in Ptolemy's catalogue: the Ram faces in the opposite direction; the Twins are portrayed standing upright; the Virgin also is depicted as standing; the Waterman faces in the contrary direction; and the Fishes both swim towards the West. The remaining figures, however, are almost precisely the same as those familiar from modern star maps. It would be surprising if *all* the figures were the same as Ptolemy's in view of his statement: "The descriptions which we have applied

---

1. Cf. Neugebauer-Parker, EAT III, pp72–74. See cover design.
2. Ibid., pp62–64.
3. Exaltation is a translation of the Greek term *hypsoma*, which is the place in the zodiac (in later astrology indicated as degrees in a particular zodiacal sign) where a planet was said to exert its strongest influence; cf. Rochberg-Halton (1988), pp56–57, who indicates a possible connection between hypsoma and the Babylonian term *asar nisirti* or *bīt nisirti*, meaning 'place of secret' or, according to Weidner (1967), pp39–40, 'place of secret revelation'.
4. Hunger-Pingree (1999), p28. Cf. also below, p116, n4.

to the individual stars as parts of the constellation are not in every case the same as those of our predecessors. . . ."[1] On the whole, however, it is impressive just how similar the modern constellation figures are to the Egyptian representations of the sidereal signs.

The rectangular zodiac from the Temple of Khnum, about 2½ miles northwest of Esna, is reproduced as one of the Plates in *Egyptian Astronomical Texts*, although the signs of Virgo, Libra and Scorpio are missing.[2] The remaining figures—depicted on the two rectangular strips which were found on the ceiling of the temple—are virtually the same as those of the Dendera zodiac.[3] The Ram in the Esna zodiac is shown facing in the customary direction, whilst the Bull faces in the contrary direction, and so does the Crab. As in the zodiac of Dendera, the Esna zodiac portrays the planets in their signs of exaltation, again indicating Babylonian influence upon the Egyptian conception of the zodiac—as referred to above in connection with the Dendera zodiac.

Although the pictorial nature of the zodiacal divisions employed by the Egyptians of the Hellenistic and Roman periods is known from the above-mentioned monuments and also from astronomical representations belonging to other Egyptian monuments, there is a shortage of scientific, numerical information concerning the divisions of the Egyptian zodiac.

> Egypt provides us with the exceptional case of a highly sophisticated civilization which flourished for many centuries without making a single contribution to the development of the exact sciences.[4]

The Egyptians, prior to the Hellenistic and Roman periods, did not cultivate mathematical astronomy as the Babylonians did (or if they did, no trace of it remains). Whereas the Babylonians left tens of thousands of cuneiform texts from which an accurate knowledge of Babylonian mathematical astronomy has been recovered, the ancient Egyptians left very little in the way of texts. It was during the Hellenistic period, after the founding of Alexandria in 332 BC, that a scientific element become introduced into the Egyptian culture. From the Hellenistic time on, zodiacs appear in temples and tombs. Prior to this there is no trace of the twelve-fold division of the zodiac to be found in Egypt.[5] Since a well-defined coordinate system is a necessary prerequisite for mathematical astronomy, it is not surprising that a scientific approach to astronomy was lacking prior to the introduction of the zodiac from Mesopotamia.

---

1. Ptolemy, *Almagest* VII, 4 (trsl. Toomer, p340).
2. Neugebauer-Parker, EAT III (Plates), plate 29.
3. Ibid., plate 35.
4. Neugebauer, HAMA II, p559.
5. Ibid., p565.

Beginning with the Hellenistic period the rapid development of Greek astrology is also reflected in Egyptian documents. From the time around the beginning of our era we have demotic horoscopes and demotic planetary tables, all close parallels to Greek texts. We also have demotic tables which use the sexagesimal number system and apply Babylonian methods. It is clear that these texts can no longer be considered as belonging to "Egyptian" astronomy although they were undoubtedly written by Egyptians. The methods, however, are purely Hellenistic and have no connection with the Egyptian past.[1]

Belonging to this latter class of texts, written during the Roman period, are two demotic planetary Tables from which the 'Egyptian' sidereal division of the zodiac can be extrapolated. The texts are ephemerides, i.e. tables of planetary positions computed for regular time intervals. One is a large demotic papyrus—P 8279 housed in the Berlin Museum—that refers to the years from 18/17 BC to AD 12;[2] and the other, known as the 'Stobart tablets'—four wooden tablets from Thebes and now in the Free Public Museum at Liverpool—refers to the period from AD 70 to 133, although some of the years are missing.[3] The first methodical analysis of these texts was published in 1942.[4] From this it emerged that both sets of ephemerides employ a sidereal division of the zodiac into twelve equal fixed-star signs.

> [It is] very probable that our texts used a fixed division of the ecliptic, corresponding to the longitude of about $\lambda = -4°$ in the time of Augustus.[5]

That is, the Egyptian sidereal division located the vernal point at about 4° in the sidereal sign of Aries in the year 0.[6] This corresponds more or less exactly with the Babylonian sidereal zodiac.[7] In conclusion, then, the 'Egyptian' zodiac is none other than that of the Babylonians, introduced into Egypt during the Hellenistic period not later than the date of building the Temple of

---

1. Ibid.
2. Neugebauer-Parker, EAT III, pp225–240, plates 66–73. The demotic script was a simplified kind of writing for the people (*demos*) as opposed to the hieratic (priestly) hieroglyphics. It was in use in Egypt from about 700 BC to around AD 500.
3. Ibid., p243.
4. Neugebauer (1942). Cf. also van der Waerden, SA II, pp308–325 for an extensive analysis of the texts.
5. Neugebauer (1942), p243.
6. Neugebauer later emended this to between 4° and 2° of the sidereal sign of Aries; cf. Neugebauer-Parker, EAT III, p226: "It must be remembered that the Egyptian texts use sidereal coordinates, not tropical longitudes, such that the beginning of Aries should be located between about 356° and 358° of the modern tables."
7. Appendix I. Cf. also Kugler, SSB i, p121, p173 and SSB II, p513 ff., where Kugler shows that the Babylonian planetary texts belonging to the last two centuries BC locate the vernal point at about 5° of the sidereal sign of Aries; cf. also Huber (1958), who locates the vernal point at 4°28' Aries in the Babylonian sidereal zodiac in the year –100.

Khnum (ca. 246–180 BC) with the Esna zodiac. There is, however, a further interesting aspect to consider with regard to the introduction of the Babylonian zodiac into Egypt.

THE DECANS

The use in Egyptian astronomy of a sidereal zodiac comprising twelve 30° fixed-star signs belongs to the very latest period of the ancient Egyptian culture. Long before this, as is evident from inscriptions found on surviving monuments of ancient Egypt, the astronomer-priests utilized a unique sidereal division in which the stars were grouped into 36 *decans*. For our present-day knowledge of the Egyptian decans we are indebted to Neugebauer and the Egyptologist R.A. Parker, who collaborated for many years to collect and analyze all existing astronomical texts belonging to the Egyptian civilization. They published the texts and their associated commentaries in three volumes during the 1960s.[1]

> The texts with which we do concern ourselves can be associated with three major periods of development. The earliest, accessible to us in texts from the Ninth Dynasty on, attempts to measure time by the risings, later by the transits, of stars and constellations, the so-called decans. The second period, represented only by Ramesside monuments, introduces a new scheme for transits. Finally, in the Ptolemaic period, foreign influence becomes dominant with the introduction of the zodiac, to which the decans were assimilated.[2]

From the earliest period, around 2150 BC, the decans first appear on coffin lids. Their later modified use in the second period is represented by the inscriptions in the cenotaph of Seti I and the tomb of Ramses IV (about 1300 and 1150 BC).[3] Finally the Esna zodiac (ca. 200 BC) and the Dendera zodiac (pre-30 BC), already mentioned, represent the 36 decans in relation to the zodiac, showing the planets in their sign of exaltation (originally specified by the Babylonians—see page 116, n4). The Esna and Dendera zodiacs belong to the Ptolemaic period, which commenced with the conquest of Egypt by Alexander the Great in the year 332 BC. During this Hellenistic period the cultural life of Greece and Mesopotamia became infused into Egyptian culture, where the city of Alexandria founded by Alexander became a leading cultural center in the civilized world of antiquity.

Prior to their assimilation into the zodiac, the decans were 36 constellations

---

1. Neugebauer-Parker, EAT I, II, III. Cf. also Boll's valuable work *Sphaera* and the extensive text by Gundel, *Dekane und Dekansternbilder*.
2. Neugebauer-Parker, EAT I, Intr. VII.
3. Ibid., p36 ff.

88 HISTORY OF THE ZODIAC

belonging to a belt of stars running parallel to the ecliptic, but lying to the south.[1] Since there are some fifty different lists of decans surviving from ancient Egypt, there is considerable difficulty in identifying the decan constellations. However, the identity of one decan is certain, and on this basis an attempt is made to begin to identify the decans.

> That śpd and śpdt 'Sothis' are both identified with Sirius is one of the rare certainties in Egyptian astronomy.[2]

The decan Sothis is identical with Sirius (α Canis Maioris). From a late papyrus belonging to the period when Egypt was under Roman rule, an idea can be gained of the Egyptian conception of the decans, taking the decan of Sothis as the primary illustration for all the decans. Papyrus Carlsberg I, which probably belongs to the second century AD, is now in the collection of

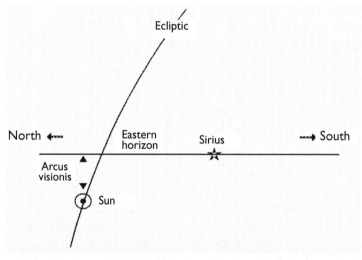

Figure 15

Heliacal Rising of Sirius (Sothis)

the University of Copenhagen, but originally it was probably from the Fayum.[3] It gives a dramatic account of the sun's movement during the day and the year in relation to the decans.

> All the stars begin in the sky ... when Sothis rises.[4]

1. Called the decanal belt.
2. Neugebauer-Parker, EAT I, p25.
3. Ibid., pp37 ff.
4. Ibid., p54.

This indicates the association of the commencement of the year with the heliacal rising of Sirius, i.e. when Sirius first appears on the eastern horizon, rising out of the rays of the dawn sun (Figure 15).

The arcus visionis of a star is the angle beneath the horizon of the sun in order for the star to become visible on the horizon. For Sirius the arcus visionis is about 10° but for most stars it is generally about 15°.

The observer who looks towards the east at dawn may perceive that certain stars lying close to the horizon appear to emerge out of the rays of the rising sun. As the sun progresses along the ecliptic from day to day so do new stars rise ahead of the dawn sun, called the *heliacal rising* of these stars. For the Egyptians the greatest event of the year was when the star Sirius first appeared on the eastern horizon at dawn, as though born out of the rays of the rising sun. This event coincided with the commencement of the Nile flood upon which the agricultural life of Egypt depended. The heliacal rising of Sirius was thus associated with the beginning of the agricultural year.

> Sirius-Sothis is the original example for the use of stars as indicators of time. With the heliacal rising of Sirius-Sothis is related the return of the flood and thus the beginning of the agricultural year. As its re-appearance indicated the ideal beginning of the year, so the heliacal risings of subsequent stars were associated with the subsequent decades.[1]

This is referred to in the text of Papyrus Carlsberg I as follows:

> [A] star dies and a star lives every decade [of days]....[2]

This statement signifies that a new 'star', i.e. a new *decan constellation* identified by a star or group of stars lying south of the ecliptic, is born on the eastern horizon (heliacal rising) every decade of days—every ten days. Simultaneously a 'star' (decan constellation) disappears from view on the western horizon and dies. Approximately every ten days a new decan is born in the east and a decan dies in the west, to spend a period of time in the *Duat*, the Egyptian term for underworld, also known as the *house of Geb*.

> According to the papyrus Carlsberg I ... every decan is invisible for 70 days between evening setting and morning rising. The star remains "in the underworld, in the house of Geb" for 70 days. There "it purifies itself and rises on the horizon like Sothis." It seems that Sirius, which indeed remains invisible for about 70 days, was taken as a model for all decans.[3]

---

1. Ibid., p107.
2. Ibid., p56.
3. Van der Waerden, SA II, p18.

Sirius-Sothis was the model decan for the 36 decans. The course of the year was determined by the successive heliacal risings of a new decan roughly every ten days so that the Egyptian year was divided approximately into 36 ten-day periods, each of which was related to the decan currently rising at dawn.

The twelve months is one way of speaking of the 36 stars.[1]

Each decan was visible during the night apart from a period of about 70 days spent in the *Duat* (*house of Geb*) after which time the decan was seen to rise again, reborn.

> [These] — that is to say, Orion and Sothis, who are the first of the gods —
> ... they customarily spend 70 days in the Duat (and they rise) again....[2]

In the decan lists the name of each decan is generally accompanied by a number of stars and is usually associated with a deity. The decans relating to the 'first of the gods', i.e. the decans corresponding to Orion and Sirius, were associated with the deities Osiris and Isis. This is known from an inscription found on the rectangular zodiac from the Temple of Hathor at Dendera (AD 14–37).

> Orion is called 'the divine and august soul of Osiris'... Sothis, the great ... daughter of Re, Isis, lady of the sky, who rises yearly in order to open a good year, who travels peacefully after her brother, the god who is Orion; her son Horus is king forever.[3]

In both Dendera zodiacs Orion is depicted holding a scepter before him and Sothis is portrayed as a recumbent cow with a star, presumably Sirius, between the horns.[4] Also both zodiacs place a falcon on a pillar separating Orion and Sothis. This is a depiction of 'Horus-who-is-on-his-pillar'. Neugebauer was unable to find any astronomical association for the falcon (Horus-on-his-pillar)[5] but possibly this figure was associated with Canis Minor, whose most prominent star is Procyon (α Canis Minoris). The Greek name *Pro-Kynos* means 'the one before the hound' (*Pro* = before, *Kynos* = hound), i.e. the one rising before Sirius. This identification accords with the depiction of Horus-on-his-pillar preceding Sothis and following Orion, since Procyon rises heliacally after Orion and before Sirius.

1. Neugebauer-Parker, EAT I, p 57.
2. Ibid., p 74.
3. Ibid., III, pp 80–81.
4. Ibid., plates 35 and 42.
5. Ibid., p 200.

THE DECANS IN RELATION TO THE ZODIAC

Major problems in the identification of the decan constellations arise when the relation of the decans to the zodiac is considered. The decanal belt lies south of the ecliptic, hence the belt of 36 decan constellations runs parallel to but south of the belt of the twelve constellations or sidereal signs comprising the zodiac (Figure 16).

Figure 16
The Ecliptic and the Decanal Belt

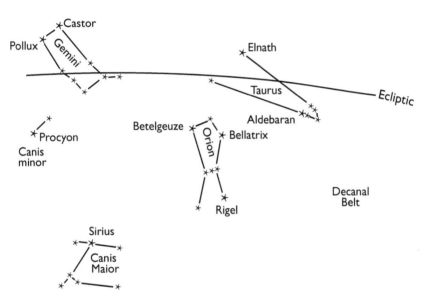

Before attempting to relate the decans to the zodiacal signs, the term *decan* requires closer definition.

> In Hellenistic and medieval astrology *decans* are defined as thirds of zodiacal signs, i.e., segments of the ecliptic of exactly 10° length. Usually these segments are not given special names but they are simply counted as 1st, 2nd, or 3rd decan of the zodiacal sign in question. In the texts discussed in this volume we are dealing with decans of a quite different character, though there can be no doubt that the Hellenistic concept is the final outcome of a development, the initial phase of which is the main theme of this study. The decans in our present sources are stars or constellations which serve by their risings or, later on, by their transits, to indicate hours of the night. They are decans in the sense of the Hellenistic terminology in so far as they are related to the 36 decades or 10-day intervals of the 12 months of the Egyptian civil calendar....[1]

1. Ibid., I, pp 95–96.

The line of development that occurred is evidently that the decans of Hellenistic astrology are the result of the assimilation of the Egyptian decans into the zodiac. This must have occurred after the introduction of the zodiac into Egypt during the Hellenistic period.

> When the Babylonian zodiac was introduced during early Hellenistic times into Egyptian celestial iconography, these 36 decanal constellations are simply located as 36 ten-degree sections in the zodiac. These, then, are the "decans"....
> Whether or not the decans actually survived into the Graeco-Roman period as instruments for telling the hours of the night, there can be no question about the assumption by them of a new role, that of identifying parts of the zodiac. This latter was the final outcome of Babylonian attempts to describe the motions of the planets and the moon and sun with respect to constellations which they meet in orbiting the sky. By the Hellenistic period the twelve zodiacal signs had become a well-defined coordinate system in Babylonian mathematical astronomy, on which the computations of Hellenistic astrology were based. One may conjecture that, when these doctrines reached Egypt, the native priests claimed to have had from time immemorial the equivalent of the zodiacal belt in the form of the belt of decans, an identification which was so much the easier to make since the zodiacal signs had still no more than a schematic association with actual constellations. Thus the decans, in their last, Hellenistic phase, became simply 10° sections of the zodiacal signs.[1]

Since the Egyptian decans are too vaguely defined to be identified with actual stellar configurations, the problem of identifying the decans can be approached by utilizing the Hellenistic concept of the decans as thirty-six 10° segments of the zodiac. But this can only be attempted if the relationship between the original decans and the 10° zodiacal sections is known. The first representation of this relationship is found on the earliest known Egyptian zodiac, the zodiac of Esna.

> The earliest zodiac known to us in Egypt is that of Esna A, now destroyed but to be dated to about 200 BC. This monument is also the first one to document an incontrovertible relationship between the zodiac and the decan lists....[2]

However, although the Esna zodiac portrays a clear association of three decans to each zodiacal sign, there are two different decan lists. The upper registers have decanal figures along with mythological figures from one list and the lower registers have decanal figures from another list.[3] It has not yet been possible to identify the decanal figures on the upper registers in the region of

---

1. Ibid., III, p168.
2. Ibid.
3. Ibid., p64 and plate 29.

Cancer, but the reconstruction of the decan list given by the upper registers can be made with certainty on the basis of the sure identification of the decan figures associated with the remaining signs of the zodiac.[1] Certain of the correctness of the equations between decans and zodiacal signs, the decan list of the upper register yields *Sopdet* (Sothis) as the first decan of Cancer.[2] This agrees with the Greek list of decanal names given by Hephaestion of Thebes (fourth century AD), who calls the first decan of Cancer Sothis.[3] The second decan list from the Esna zodiac, that of the lower registers, does not include Sothis as a decan. But in this list the decan *Knumet* is associated with the third decan of the sign of Cancer.[4] *Knumet* is also the third decan of Cancer in the list compiled form the upper registers. The Greek name for *Knumet* is Knoumis. Hephaestion's list makes Knoumis the third decan of Cancer, in agreement with both decan lists from the Esna zodiac.[5] From a detailed comparison of the names of each Egyptian decan in the two Esna lists with Hephaestion's Greek decanal names, it emerges that 24 names in Hephaestion's list agree with the Egyptian prototype names in the 'Sothis' list compiled from the upper registers of the Esna zodiac, whilst the remaining twelve names seem to be taken arbitrarily from the second list, that of the lower registers (see Table 5).[6]

Table 5
Esna Decan Lists

DECANS		FIRST LIST (UPPER REGISTERS)	SECOND LIST (LOWER REGISTERS)
Cancer	1	*Sopdet (Sothis)	Waret
	2	*Setu	Pehwey-hery
	3	*Knumet	*Knumet
Leo	1	*Khery-kheped-Knumet	*Khery-kheped-Knumet
	2	*Het-djat	*Het-djat
	3	*Pehwey-djat	Djat
Virgo	1	*Tjemat	Pehwey-djat
	2	*Weshaty-Bekaty	Tjemat
	3	*Ipsed	Weshaty

1. Ibid., p168 ff.
2. Ibid., p170.
3. Ibid.
4. Ibid.
5. Ibid., p169.
6. Ibid., pp170–171.

94 HISTORY OF THE ZODIAC

DECANS		FIRST LIST (UPPER REGISTERS)	SECOND LIST (LOWER REGISTERS)
Libra	1	*Sebshesen	Bekaty
	2	*Tepi-a Khentet	Ipsed
	3	*Khentet-hert	Sebshesen
Scorpio	1	Khentet-Khert	Tepi-a Khentet
	2	Themes-en-Khentet	Hery-yeb wia
	3	Sapty-khenwey	Sapty-khenwey
Sagittarius	1	*Her-ab-uaa	Seshmu
	2	*Seshmu	Sawey Seshmu
	3	*Kenemu	*Kenemu
Capricorn	1	Tepi-a Semed	Tepi-a Semed
	2	Semed	Pa-sebu-wity
	3	Seret	Semed
Aquarius	1	Sawey Seret	Seret
	2	Khery-kheped-Seret	Sawey Seret
	3	Tepi-a Akhwey	Tepi-a Akhwey
Pisces	1	Akhwey	Akhwey
	2	Tepi-a Bawey	Tepi-a Bawey
	3	*Bawey	*Bawey
Aries	1	*Khentu-heriyew	*Khentu-heriyew
	2	*Khentu-kheriyew	*Khentu-kheriyew
	3	*Sawey-Qed	Qed
Taurus	1	*Khau	Sawey-Qed
	2	*Aret	Khau
	3	*Remen-hery	Aret
Gemini	1	*Tjes-areq	Remen-hery
	2	*Waret	Tjes-areq
	3	Tepi-a Sopdet	Remen-khery

*Denotes a decan in Hephaestion's list

Clearly there is an overall agreement between Hephaestion's list and the 'Sothis list' compiled from the upper registers of the Esna zodiac, with 24 equations between the Greek decanal names and their Egyptian prototypes. This 'Sothis list' has a long history, appearing first in the time of Seti I (1303–1290 BC), and as late as the time of Trajan (AD 98–118).[1] In all probability it was based on astronomical observations (unlike some decan lists).

The zodiac of Esna (ca. 200 BC) gives the first expression of a relationship between the decans and the zodiac, albeit in two different ways. Some 160 years later the circular zodiac of Dendera also portrays the 36 decans, around the outside of the twelve zodiacal signs.[2] The relationship between the decans

1. Ibid., pp 133 ff.
2. Ibid., plate 35.

and the zodiacal signs in the Dendera zodiac is apparently the same as that in the second list (from the lower registers) in the Esna zodiac.[1] Certainly it is the same decan list, one which does not include Sothis as a decan. However, because of the irregular arrangement of the decans around the zodiac circle the alignment of *Knumet* (Knoumis) with the third decan of Cancer as in the Esna zodiac, although quite feasible, is not certain—and similarly with the alignments of the remaining decans.

There is no single consistent relationship between the Hellenistic conception of the decans as 10° segments of the ecliptic and the Egyptian decans as constellations in the decanal belt lying south of the ecliptic. The Dendera zodiac gives one relationship (not including the decan Sothis) and this appears to be the same as the relationship found in the decan list from the lower registers of the Esna zodiac. On the other hand the decan list (including Sothis) compiled from the upper registers of the Esna zodiac gives a different relationship, one that is in overall agreement with the Greek list of decanal names given by Hephaestion of Thebes. In the reconstruction of the Egyptian version the Sothis decan is related to the first 10° segment of the sign of Cancer. This agrees with Hephaestion, who names the first decan of Cancer 'Sothis'. Altogether there are 24 equations between the Greek names in Hephaestion's decan list and the Egyptian prototype names in the 'Sothis list' from the Esna zodiac (Table 5).

Evidently it is impractical to attempt to identify the Egyptian decans by utilizing their relationship to 10° sections of the zodiac since there is no consistent relationship to be found. Nor is it known exactly what the basis of any such correspondence is between the Egyptian decans and the thirty-six 10° segments of the ecliptic.[2] Moreover, the use of decans by Egyptian priests was a qualitative astronomical scheme rather than a precise numerical procedure.[3] Therefore any correspondence between the Egyptian decans and 10° sections of the ecliptic can only approximate rather than represent an accurate astronomical relationship.

---

1. Ibid., p72.
2. It is generally assumed that the Greeks identified the decans with their paranatellonta in the zodiac. 'Paranatellonta' are stars that rise simultaneously; cf. Neugebauer-Parker, EAT I, p97, n2, and van der Waerden, SA II, pp18–19. If this assumption is correct, then according to Hephaestion's list the stars at the beginning of Cancer should rise at the same time as the star Sirius. In fact, for the latitude of Esna, computation shows that it was the star cluster Praesepe (the Beehive) near the middle of Cancer (at 12½° Cancer according to Appendix I) that rose over the horizon with Sirius around 200 BC.
3. Neugebauer-Parker, EAT I, p95.

5

THE SIDEREAL ZODIAC IN MESOPOTAMIA

FROM THE WEALTH OF AVAILABLE EVIDENCE it is clear that early Egyptian astronomers used a division of the stars into 36 decans, constellations lying to the south of the ecliptic. This division was utilized as the basis for a yearly calendar comprising approximately 36 ten-day periods, and as a scheme for recording the passage of hours at night.[1]

Further to the east, in the contemporary culture that flourished on the banks of the Tigris and Euphrates, Babylonian astronomers developed a quite different approach to astronomy. Instead of remaining satisfied with general qualitative astronomical schemes they sought to develop accurate numerical procedures. Their attention was directed primarily to the irregular movements of the planets along the belt of stars known as the zodiac, through the middle of which runs the ecliptic. Because they were concerned not only with observation of these movements but also with prediction of the recurrence of planetary phenomena, their astronomy inevitably progressed beyond simple descriptions of the phenomena to the determination of arithmetical expressions for computing planetary motions.[2] Thus the Babylonians were responsible for the primary development of science, for they were the first to combine observation with measurement and mathematical reckoning.

> Nowhere within ancient civilizations known to us did the sciences originate independently, neither in pre-Hellenic nor in early Greek civilization, in the ancient Near East, on the Iranian plateau, nor in pre-Arian or Arian India—with the sole exception of Mesopotamia, probably in the early second millennium. It is at this single center that abstract mathematical thought first appeared, affecting, centuries later, neighbouring civilizations, and finally spreading like a contagious disease.[3]

1. Neugebauer-Parker, EAT I, pp 95 ff.
2. Aaboe (1974) and Sachs (1974) discuss Babylonian observational and scientific astronomy.
3. Neugebauer, HAMA II, p 559.

Irrefutable proof of the Babylonian contribution to the development of astronomy was supplied by the discovery from cuneiform sources of a parameter for the lunar synodic month identical with the parameter used by Hipparchus.[1] Subsequent research has shown that many of Hipparchus' fundamental parameters are Babylonian in origin[2] and, as far as is known, Hipparchus was the first to introduce Babylonian parameters into Greek mathematical astronomy.[3] The geometric-cinematic models of planetary motion developed by the Greek astronomer Apollonius[4] became associated through Hipparchus with accurate numerical parameters, some of which were Babylonian and others derived from Hipparchus' own observations.

Somehow Hipparchus inherited the end results of Babylonian astronomy.[5] It is unlikely that he knew the intricacies of the Babylonian computations of ephemerides,[6] but he was nevertheless able to extract from Babylonian astronomy the parameters that he needed. He thus achieved a union of the Greek geometrical approach to astronomy with the arithmetical approach of the Babylonians. If this convergence of the two 'schools' of astronomy had not taken place, it is difficult to envisage how a work like the *Almagest* could have been written. Thus the development of Babylonian arithmetical procedures for computing planetary motions was crucial to the foundation of mathematical astronomy. It is sufficient to mention the use of the zodiac in astronomy and the continued use of the sexagesimal number system in the division of the circle and in time-reckoning to realize that Babylonian influence is present in the very foundations of the science of astronomy.

NORMAL STARS

When did astronomy start to become scientific? Or, since any science depends initially upon observations, at what point in time was the practice of making accurate observations of the stars instituted?

1. Kugler, *Die Babylonische Mondrechnung*, p111.
2. Aaboe (1955); cf. also Toomer (1988).
3. Neugebauer, HAMA I, p309; cf. also van der Waerden, SA II, p298 for the possibility of a source prior to Hipparchus.
4. Neugebauer (1959); cf. van der Waerden (1974) for a discussion of the possible Pythagorean origin of the epicycle theory which, however, is generally associated with Apollonius of Perga (third century BC), whose geometric-cinematic model involves a deferent circle centered on the earth and rotating eastward (counter-clockwise) at constant speed together with an epicycle circle (centered on a point on the circumference of the deferent circle) upon which the planet rotates counter clockwise at constant speed.
5. Neugebauer, HAMA I, pp 341–342 discusses the problems associated with this transmission.
6. Ibid., p330.

[A] continuous record of dated observations began with the reign of Nabonassar, whose first year was our 747 BC, "from which date," says Ptolemy, "we possess the ancient observations continued practically to the present day."[1]

Throughout the *Almagest* Ptolemy uses the beginning of the reign of Nabonassar as his epoch date, i.e. the starting point from which his computations are based. He states this to be noon Alexandria time, on the first day of the Egyptian month Thoth in year one of Nabonassar's reign, which corresponds to mid-day, 26th February, 747 BC.[2] Ptolemy's statement concerning "a continuous record of dated observations" has been substantially confirmed by data from cuneiform sources.

> [Eclipse] texts ... are more uniformly distributed over the whole time beginning in the middle of the 8th century BC, in perfect agreement with Ptolemy's statement that the ancient observations were almost completely preserved from the time of Nabonassar on.[3]

Since all science is founded on the principle of observation (followed by analysis), the introduction of the scientific principle in astronomy can be traced back to the start of the era of Nabonassar, from which time dated observations commenced.

These early observations were of eclipses. Then observations of planetary positions began to be collected. The method adopted by the Babylonians was to record the position of a planet in relation to a nearby fixed star. These reference stars all lie within a zone extending between 10° north and 7½° south of the ecliptic, and are known as 'Normal Stars'.[4] The use of Normal Stars in the astronomical observations of the Babylonians is attested in cuneiform sources early in the period of scientific observation.

> Normal Stars are known to us from the Diaries which go back to the seventh century and, for the Seleucid period, from the Goal-Year texts and from the Normal-Star Almanacs. In all these texts, positions of planets at given dates are recorded in relation to a nearby Normal Star. In the Diaries the same also holds for the moon. Although much of these data is not directly observed but computed, we have here an enormous collection of ultimately empirical data.[5]

---

1. Heath, *Greek Astronomy*, intr.
2. Neugebauer, HAMA I, p59.
3. Neugebauer, HAMA I, p352.
4. Terminology introduced by Epping, *Astronomisches aus Babylon*, p115; cf. Sachs (1974), p 46 and Hunger-Pingree (1999), pp148–149 for a list of Normal Stars.
5. Neugebauer, HAMA I, p546.

The terms 'Diaries', 'Goal-Year texts', and 'Normal-Star Almanacs' constitute a nomenclature for classifying various kinds of cuneiform texts.[1] The Diaries contain observational data; the Goal-Year texts were used to predict the recurrence of planetary phenomena on the basis of certain planetary periods, e.g. the 8-year period for Venus after which time Venus returns to the same Normal Star and is simultaneously at the same phase in relation to the sun; and the Normal-Star Almanacs which provide computed information about the entrance of planets into zodiacal signs, the occurrence of solstices and equinoxes, etc.

As might be expected, the analytical approach to astronomy, in which information about planetary positions is derived computationally rather than through observation, is more strongly evident in the later stage of Babylonian astronomy, especially during the Seleucid era.[2] It was towards the end of the Persian (Achaemenid) reign or at the beginning of the Seleucid era that the mathematical astronomy of the Babylonians emerged as a fully fledged science comprising linear arithmetical schemes for computing the phenomena of the moon and each of the planets. In this later phase of mathematical astronomy the zodiac is used consistently as the coordinate system for computational purposes alongside the continued use of Normal Stars for observational purposes. The transition from observational astronomy to mathematical astronomy is thus characterized by the adoption of the zodiac as the primary frame of reference.

Babylonian astronomy displays a three-fold path of development: an early, *primitive* stage; a second, *observational* phase, beginning in the middle of the eighth century BC when astronomical observations were collected; and a subsequent period of *mathematical* astronomy coinciding with the Seleucid era (having developed over a period of about 150 years preceding the Seleucid era).[3] The steps in this development are difficult to trace, since the cuneiform sources provide only the completed mathematical-astronomical procedures and no accompanying theory to explain how they were derived.

> [We] have practically no concept of the arguments, mathematical as well as astronomical, which guided the inventors of these procedures.... Hence we are very far from any 'history' of Babylonian astronomy and must be satisfied to

---

1. Introduced by Sachs, cf. Sachs (1948) and Pinches-Sachs (1955).
2. Beginning in April 311 BC; cf. Neugebauer, ACT I, p32.
3. The start of the development of Babylonian mathematical astronomy coincided more or less with the first recorded use of the sidereal zodiac, cf. Britton (1999), pp244–246. Further, these three phases, although holding generally, are not rigid time divisions, and there are examples of exceptions, one being that the earliest known astronomical observations (the Venus tablets of Ammi-saduqa) date back to about 1700 BC; cf. Hunger-Pingree (1999), pp32–34.

accept it as a completed system of admirable elegance and efficiency but without really understanding its development.[1]

However, it may be conjectured that when the sidereal zodiac acquired definition in relation to the Normal Stars the basis was laid for the transition from observational to mathematical astronomy. It is this definition of the Babylonian sidereal zodiac in relation to the Normal Stars that is crucial for the history of the zodiac and is therefore central for this thesis. Since the first record of the zodiac belongs to the first half of the fifth century BC,[2] the transitional development from observational to mathematical astronomy must have begun at this date, if not earlier. Britton argues that the system of measuring longitudes in terms of the twelve signs of the Babylonian sidereal zodiac was introduced between −463 and −453.[3]

NORMAL STARS IN RELATION TO THE SIGNS OF THE ZODIAC

The definition of the zodiac was an important step in the development of Babylonian astronomy. It seems to have occurred during the first half of the fifth century BC and eventually superseded the use of Normal Stars, although the two coordinate systems continued to be used side by side throughout the whole Seleucid period.

> The 'Diaries', the 'Goal-Year texts', and the 'Normal-Star Almanacs' describe the positions of the planets and of the moon with respect to a set of 31 reference stars, called 'Normal Stars', in Epping's terminology. The ecliptic with its division into 12 signs is present in these texts but the positions of the celestial bodies are related to the Normal Stars not by coordinates counted in degrees but by distances measured in 'cubits' and 'fingers'.[4]

The simultaneous appearance of the two coordinate systems in the same texts leaves no room for doubt that the signs of the zodiac were defined in relation to Normal Stars. The origin of this definition, however, is unknown. Curiously, only one explicit reference to the definition of the zodiac in terms of Normal Stars has been unearthed so far.[5] Although only a fragment, it is

---

1. Neugebauer, HAMA I, p348.
2. In the list of solar eclipses preserved for the years from 475 to 457 BC; cf. Aaboe-Sachs (1969), p17.
3. Britton (1999), p244 indicates that the twelve signs of the Babylonian sidereal zodiac "appeared between −463 and −453."
4. Neugebauer, HAMA I, p545.
5. Sachs (1952, 2); cf. also Hunger-Pingree (1999), pp150–151.

the earliest known star catalogue and it assigns longitudes of stars as degrees in the signs of the Babylonian sidereal zodiac. The style of the catalogue led Sachs to conclude that its probable date of composition lies before the beginning of the Seleucid era, i.e. in the Persian (Achaemenid) period.[1]

A careful analysis by Huber of the longitudes of stars recorded in the Babylonian star catalogue,[2] together with other information on star longitudes gathered from cuneiform sources, was published in 1958.[3] His procedure was to tabulate Normal Star positions as longitudes in the twelve signs of the zodiac for eleven occasions (during the first two centuries BC) on which cuneiform texts give planetary positions in both systems simultaneously. This comparison, including star longitudes from the Babylonian star catalogue, enabled him to determine within fairly precise limits the relationship between the Babylonian zodiac and the location of the vernal point in −100. His analysis led to the conclusion that in −100 the vernal point was located at 4°28' in the sign of Aries, according to the Babylonian definition, with an error interval of ±20'.[4] This result enables the Babylonian zodiac to be reconstructed quite accurately, using stellar longitudes computed from modern star catalogues. Conveniently, a tabulation of stellar longitudes and latitudes for all stars listed in Ptolemy's catalogue was published by Peters and Knobel,[5] with epoch date AD 100, close (relatively speaking) to Huber's date of -100. The addition of 1°40' to each of the longitudes in the catalogue of Peters and Knobel is a simple way to reconstruct the Babylonian star catalogue in its entirety.[6] By adding this value, the two first magnitude Normal Stars Aldebaran and Antares are placed at 15° Taurus and 15° Scorpio, respectively, which agrees with statements made by some Greek astrologers:

> Cleomedes states (*De motu* I, 11 p106, 25 to 108, 5 Ziegler) that there exist two bright stars such that the rising of one coincides with the setting of the other:

---

1. Ibid., p150. The Achaemenid Dynasty ruled in Babylonia from −537 to −329.
2. Only fragments of this Babylonian star catalogue remain.
3. Huber (1958) used eleven observations from the first two centuries BC where the two systems (Normal Stars and sidereal zodiac) were used simultaneously.
4. Ibid., p208; cf. also van der Waerden (1953), p222 for an earlier reckoning, where the vernal point for the year -100 is located in the vicinity of 4°18' Aries.
5. Peters-Knobel, *Ptolemy's Catalogue of Stars*, pp58 ff.
6. According to Huber's result the vernal point lay at 4°28' Aries ± 20' in −100. Two hundred years later, in AD 100, the vernal point having traveled 2°47' (allowing a rate of precession of 1° in 72 years) lay at 1°41' Aries ± 20'. Since the longitudes computed by Peters and Knobel are measured from the vernal point of AD 100, approximately 1°40' must be added to their computed values in order to obtain the corresponding values in the Babylonian star catalogue. However, in Appendix I the epoch of the Babylonian star catalogue is -100 rather than AD 100, and the stellar longitudes are taken not from the catalogue of Peters and Knobel but from the catalogue compiled by the Hipparcos satellite, the most accurate modern star catalogue.

Aldebaran (α Tauri) and Antares (α Scorpii), both being located at the 15th degree of their respective sign.[1]

The diametrically opposite positions of Aldebaran and Antares in Taurus 15° and Scorpio 15°, respectively ... is also given in a Greek treatise which goes under the name of the 'Anonymous of the Year 379'.[2]

Hephaestion of Thebes also lists the longitude of Aldebaran as Taurus 15°.[3]

The question as to why these statements drawn from Greek astrology should have any bearing on the Babylonian zodiac is discussed further below, but first let us consider Neugebauer's conclusions regarding these statements.

Neugebauer argues that the longitudes of Taurus 15° for Aldebaran and Scorpio 15° for Antares were tropical longitudes computed from Ptolemy's catalogue of stars in the *Almagest*. He concludes that Cleomedes wrote around AD 370, and the "Anonymous of the Year 379" obviously belongs to the same time. Neugebauer states:

> The longitudes given by Cleomedes are 2⅓° greater than in the *Almagest*; according to the ancient constant of precession of 1° per century the proper date for these longitudes would be 233 years after the epoch of Ptolemy's catalogue, hence 138 + 233 = 371 AD.[4]

In support of this Neugebauer adds in a footnote:

> The diametrical position of the two stars [Aldebaran and Antares] is once more mentioned by [the Greek astrologer] Rhetorius, but now for Taurus/ Scorpio 16;20° [Taurus 16⅓° and Scorpio 16⅓°], i.e. for about AD 500/510.[5]

It is clear that in the case of Rhetorius the longitudes given for Aldebaran and Antares are definitely in terms of the tropical zodiac. According to Neugebauer's line of reasoning, the longitudes given by Rhetorius are 3⅔° greater than in the *Almagest*. Allowing a rate of precession of 1° per century, the constant of precession computed by Ptolemy, 367 years have to be added to Ptolemy's epoch date, hence 138 + 367 = 505 AD, i.e. this is correct for about AD 500/510. Horoscopes drawn up by Rhetorius confirm this date.

---

1. Neugebauer, HAMA II, p960. As Neugebauer points out, "The simultaneity of the rising and setting is only fictitious. . . ." He is right regarding observers in the northern hemisphere. However, in the southern hemisphere, from below the Tropic of Capricorn, there are a few days each year when Aldebaran and Antares can be seen simultaneously; cf. K. A. Pickering (1992), p15.
2. Ibid.
3. Neugebauer-van Hoesen, *Greek Horoscopes*, p187.
4. Neugebauer, HAMA II, p960.
5. Ibid., n4. Words in brackets [ ] added by R. P.

According to Neugebauer, it is reasonable to draw a similar conclusion to the one he draws for Rhetorius in the cases of Cleomedes, Hephaestion of Thebes, and the 'Anonymous of 379', following exactly the same line of reasoning. Let us examine this more closely, beginning with Cleomedes. The only writing of Cleomedes that has been preserved is *De motu* (*On the Circular Motion of the Celestial Bodies*), and it is based to a certain extent on the (no longer extant) works of the Stoic philosopher Posidonius (died around 50 BC), who taught at Rhodes. Nowhere in *De motu* is there any reference to the works of Ptolemy, whereas Posidonius is referred to repeatedly. It seems possible, therefore, that Cleomedes' statement regarding the longitudes of Aldebaran (Taurus 15°) and Antares (Scorpio 15°) are adopted from Posidonius. Little is known about Posidonius except that he initially estimated the Earth's circumference to be 240,000 stades and later revised this figure to 180,000 stades.[1] There is also a remark made by Diogenes Laertius about "the school of Posidonius," linking Posidonius to the concept of 'steps' ($\beta\alpha\theta\mu o\iota$) used in Greek astrology.[2] And according to St. Augustine, Cicero indicated that Posidonius was "much given to astrology" and that he was "a great astrologer and philosopher."[3] It is conceivable that Posidonius, whose school existed on Rhodes, was a recipient of astrological ideas of Babylonian origin transmitted by Berossus, who—as described below—founded an astrological school on the island of Cos in the third century BC. In this case a line of transmission of knowledge of the defining axis of the Babylonian zodiac between Aldebaran (Taurus 15°) and Antares (Scorpio 15°) could be conjectured: Berossus—Posidonius—Cleomedes.

Let us now consider Hephaestion of Thebes and the 'Anonymous of the Year 379'. Hephaestion of Thebes (late fourth/early fifth century AD) was an Egyptian (or Greco-Egyptian) astrologer from Thebes, who wrote his *Apotelesmatica* in three books, the first two of which rely heavily on Ptolemy's *Tetrabiblos*.[4] It is interesting that Hephaestion in Book II, 18, probably written around AD 390,[5] also gives the longitude of Regulus as Leo 5°,[6] which agrees exactly with the sidereal longitude of Regulus given in the reconstructed Babylonian star catalogue (Appendix I). In the case of Hephaestion of Thebes

---

1. Evans (1998), p 65. The stade (*stadion*), indicating the length of a race track, was a Greek unit for measuring distance. Ancient sources vary in their estimates of the stade, with between eight and ten stades to a mile. The figure indicated by Posidonius is remarkably good, considering that the earth's actual circumference is approximately 25,000 miles.
2. Diogenes Laertius VII, 146 (Loeb II, pp 250/251); cf. Neugebauer, HAMA II, p 671.
3. St. Augustine, *De Civitate Dei* v. 2 and v. 5; cf. Tester (1987), pp 52–53 and p 113.
4. Ed. D. Pingree (1973/1974); cf. also A. Engelbrecht (1887).
5. Neugebauer-van Hoesen, *Greek Horoscopes*, p 187.
6. Ibid.

it is evident that he drew upon Ptolemy and therefore Neugebauer's line of reasoning, quoted above, probably does apply to Hephaestion of Thebes in his listing of the longitude of Aldebaran as Taurus 15°, i.e. that this longitude was obtained by computation from the *Almagest* by adding 2⅓° to the tropical longitude of Taurus 12⅔° given for Aldebaran in Ptolemy's catalogue of stars (2⅓° being approximately the amount of precession from Ptolemy's until Hephaestion's time, allowing for a rate of precession of 1° per century, as computed by Ptolemy). If this is the case, then Taurus 15° given as Aldebaran's longitude (and the same applies to Leo 5° given as the longitude of Regulus by Hephaestion) is a tropical longitude and not a sidereal longitude adopted from the Babylonian star catalogue defining the Babylonian sidereal zodiac and transmitted to Greek astrology via a line of transmission such as that indicated above for Cleomedes.

Then there is the case of the 'Anonymous of the Year 379', whose astrological text was published by Cumont in the fifth volume of *Catalogus Codicum Astrologorum Graecorum*.[1] The text concerns the use of thirty bright stars— the main stars belonging to certain constellations—for the prediction of events, utilizing (to a certain extent) astrological rules laid down by Ptolemy in the *Tetrabiblos*. The 'Anonymous of the Year 379' refers to the simultaneous rising and setting of Aldebaran (Taurus 15°) and Antares (Scorpio 15°). Not only the longitudes of Aldebaran and Antares but also most of the longitudes of the other stars given in the text, e.g. Rigel in the 23rd degree of Taurus, Sirius in the 20th degree of Gemini, Pollux in the 29th degree of Gemini, Spica in the 29th degree of Virgo, etc., agree exactly with the longitudes of these stars given in terms of the Babylonian sidereal zodiac (Appendix I). Nevertheless, from the testimony of the 'Anonymous of the Year 379' himself, it is clear that he used Ptolemy's catalogue of stars and converted the longitudes using Ptolemy's constant of precession, exactly as indicated above by Neugebauer. For the 'Anonymous of the Year 379' writes:

> As far as we are concerned, we want to preserve the memory of it for people who are worthy of the stars, and we have taken from astrological doctrine all the useful things that Ptolemy has expounded concerning this science.... To begin with, we make known to you the effective virtue of each fixed star, and we have entered in the table the degree of longitude at which each star was in the time of the consulate of Olybrius and Ausonius, that is [379], in the year in which we have composed this treatise. In truth, as divine Ptolemy has shown, during one hundred years the stars move one degree towards the parts that follow the equinoxes and solstices.[2]

1. CCAG V, 1 (pp194–211).
2. Ibid., p195.

From this it is clear that the 'Anonymous of the Year 379' drew up a table of fixed stars computed using Ptolemy's precession constant of one degree in one hundred years. Since this constant of precession is indicated in *Almagest* VII, 2 immediately preceding the catalogue of stars in Books VII and VIII, it is reasonable to assume that the 'Anonymous of the Year 379' made use of Ptolemy's catalogue of stars to derive his own table of longitudes of fixed stars. Thus Neugebauer's line of reasoning in this case is proved to be true.

However, let us return to consider the case of Cleomedes, in particular Neugebauer's conclusion that Cleomedes wrote his astronomical textbook *De motu* around AD 370. Neugebauer's line of reasoning here is based on the *a priori* assumption that because Cleomedes gave the longitude of Aldebaran as Taurus 15° and that of Antares as Scorpio 15°, he must have lived around AD 370. In other words, Neugebauer's argument is a circular one based on this *a priori* assumption. For there is an equally good case to be made that Cleomedes lived in the first century AD. Cleomedes states at the end of *De motu*: "The preceding teachings are not the author's own opinion but collected from older or more recent summaries; much of it is taken from Posidonius."[1] Here, however, it is difficult to know how long after Posidonius (died around 50 BC) Cleomedes wrote his astronomical text book. Heath came to the conclusion that Cleomedes wrote *De motu* in the first century BC, i.e. immediately after the death of Posidonius, and that, moreover, "As [Cleomedes] seems to know nothing of the works of Ptolemy, he can hardly ... have lived later than the beginning of the second century AD."[2] Based on considering the content of *De motu*, Dicks concludes that Cleomedes wrote his astronomical text book during the first century AD,[3] pinpointing a time in between that suggested by Heath, on the one hand, and that computed by Neugebauer on the other hand. If Heath or Dicks are right, then Neugebauer's argument for dating Cleomedes using the fact that he gave the longitude of Aldebaran as Taurus 15° and that of Antares as Scorpio 15° is spurious, and the only conclusion is that Cleomedes' indications concerning the longitudes of these two stars is drawn from another source (not Ptolemy), which leads us back to the conjecture that the line of transmission was: Berossus—Posidonius—Cleomedes, i.e. that for Aldebaran and Antares Cleomedes was quoting sidereal longitudes stemming from the old Babylonian star catalogue (reconstructed in Appendix I).

Now let us return to consider the question as to why the statements drawn from Greek astrology should have any bearing on the Babylonian zodiac, focusing upon Berossus as one source for the line of transmission referred to

1. Cleomedes, *De motu* (ed. Ziegler).
2. Heath (1931), p144.
3. Dicks, 'Cleomedes', DSB II, pp138–140.

here. Greek astrology is relevant here, since it is known that Greek astrologers were the direct recipients of Babylonian star lore, as is evident in the case of the astrological school founded by Berossus on the Greek island of Cos early in the third century BC.[1] From sources such as Berossus, Babylonian star lore was transmitted to Greece and became incorporated into the corpus of Greek astrology. The statements of Greek astrologers may thus offer direct insight into the nature of Babylonian astronomy, and in this instance the singling out of Aldebaran and Antares from all other stars, by virtue of their special oppositional relationship to one another, could reflect the reasoning underlying the original definition of the signs of the Babylonian zodiac as the system to replace the system of Normal Stars. It may be conjectured that at some time in the early fifth century BC some Babylonian astronomer (or group of astronomers), while making observations of the moon and planets in relation to Normal Stars, realized that two of the most prominent Normal Stars, Aldebaran and Antares, divide the zodiac exactly in half, and that the axis between them could therefore serve as the defining reference axis for all the stars of the zodiacal belt. In this way, therefore, the Normal Stars came to be related to a new system, namely the system of zodiacal signs in which the zodiacal belt was divided into twelve 30° sectors or signs, with the two signs Taurus and Scorpio defined so that Aldebaran and Antares were located at the center of these signs, respectively—Aldebaran at 15° Taurus and Antares at 15° Scorpio.[2] With this as the basic, initial definition of the structure of the Babylonian zodiac, it was then simply a matter of measuring the distance in degrees of other Normal Stars from the Aldebaran-Antares axis in order to deduce the longitudes of Normal Stars in the various signs, with Regulus at 5° Leo, Spica at 29° Virgo, etc.[3] Appendix I gives the reconstructed Babylonian sidereal zodiac based on this *intrinsic definition* provided by the fiducial axis Aldebaran (Taurus 15°)—Antares (Scorpio 15°).

With respect to the question as to why the zodiacal belt of Normal Stars was

---

1. Schnabel, *Berossos und die babylonisch-hellenistische Literatur*, pp 250–275 for Greek fragments of Berossus' writings.

2. Fagan, *Zodiacs Old and New*, p 21 lists Aldebaran at 15° Taurus and Antares at 15° Scorpio. Cf. also Gleadow, *The Origin of the Zodiac*, p 28.

3. Sachs (1952, 2), p 146. Sachs' reading for Spica from the late Babylonian star catalog is given tentatively as 28°(?) Virgo. A sidereal longitude of 28° Virgo does not greatly affect the line of reasoning presented here, since it may simply reflect a degree of inaccuracy inherent in the measurement of stellar longitudes by Babylonian astronomers. Moreover, according to van der Waerden (1953), p 227, "on the Babylonian zodiac Spica was assumed to lie between 28° and 29° Virgo." Unfortunately little is known about *how* the Babylonians measured the distances of stars from one another in degrees. As discussed below, knowing that the full moon's breadth is about ½° it is possible that distances were estimated as a number of "breadths of the moon" and converted to degrees.

divided into *twelve* signs, each 30 degrees long, van der Waerden in his early article 'History of the Zodiac' writes:

> There are twelve signs, because there are twelve months in the schematic year of MUL.APIN.... The signs were made of equal length in order to get months of equal duration; they were divided in 30 degrees each because the schematical months were supposed to contain 30 days each.[1]

In other words, in the early fifth century BC when the division of the zodiacal belt of Normal Stars into twelve signs was originally formulated, there already existed a schematic calendar devised earlier by Babylonian astronomers, known from the text MUL.APIN. The text of MUL.APIN consists of two tablets dated around 687 BC in which are listed, among many other things, the rising of stars and constellations in terms of a schematic year of twelve months each 30 days long.[2] This is a 'schematic' year, because the actual civil calendar in Babylon operated with lunar months, which fluctuate in length, being either 29 or 30 days long, and in an intercalation year (roughly every third year) there were thirteen instead of twelve lunar months. The MUL.APIN calendar scheme thus represented an idealized year: the ideal of a solar calendar (Figures 11 and 12) rather than the actual year of twelve (or thirteen) variable-length lunar months. With this scheme already in existence, the originator of the system of zodiacal signs was influenced by it in such a way as to specify a twelvefold division of the zodiacal belt into signs, each sign consisting of 30°, analogous to the twelvefold division of the year into schematic months, with each month consisting of 30 days. Once the idea of this division of the zodiacal belt, analogous to the schematic division of the year, had been formulated, it was simply a matter of defining where the signs should lie in relation to the Normal Stars comprising the zodiacal belt.

Evidently the Babylonian sidereal zodiac originated in the fifth century BC It was devised as an alternative system to that of the Normal Stars belonging to the zodiacal belt. The division of the zodiacal belt into twelve signs each 30° long was analogous to the schematic division of the year into twelve months, each 30 days long, formulated in the text MUL.APIN around 687 BC. The relationship between the Normal Stars belonging to the zodiacal belt and the division into zodiacal signs was specified by the adoption of the Aldebaran-Antares axis as the fiducial axis for the Babylonian zodiac, with Aldebaran at the middle of the sign of Taurus and Antares at the middle of the sign of Scorpio. According to van der Waerden, this

---

1. Van der Waerden (1953), p218.
2. Hunger-Pingree (1989) and (1999); cf. also Weidner (1924), pp186–208 and (1931–32), pp170–178, also Pritchett-van der Warden (1961), pp17–51, esp. pp43 ff.

[is] very plausible. The special position of Aldebaran and Antares is mentioned already in the text MUL.APIN, which says:
(1) MUL.MUL rises and GIR.TAB sets;
(2) GIR.TAB rises and MUL.MUL sets (see *Cuneiform Texts* 33, Plate 4, Rev. III, lines 13–14, or Kugler, SSB Ergänzungsband I, pp 22–23). These two lines are from the beginning of a long list of simultaneous risings and settings. So the two phenomena were considered important. MUL.MUL is the Pleiades (main star η Tauri), and GIR.TAB is Scorpio (main star Antares). According to the text, the two are at some time in one and the same horizon, and at another time again in the same horizon.... In the sixth and fifth century, when the zodiac became prominent and bright zodiacal stars were preferred, the stars Aldebaran and Antares became prominent and were used to define the midpoints of the signs Taurus and Scorpio.[1]

Once adopted as the primary reference for the new system of the twelve signs of the zodiac, the longitudes of the remaining Normal Stars were defined in terms of sign and degree in the Babylonian zodiac by determining their distances from the Aldebaran-Antares axis. The relationship between Normal Stars and the Babylonian zodiac was recorded in a star catalogue, the world's first star catalogue, thus constituting the definition of the Babylonian zodiac. In this way the transition from the system of Normal Stars to the system of zodiacal signs was accomplished, and herein lies the origin of the Babylonian zodiac.

This *intrinsic definition* of the sidereal zodiac enables it to be reconstructed exactly (Appendix I). The resulting reconstruction of the sidereal zodiac is in exact agreement with the result determined by Huber for the Babylonian definition of the zodiac,[2] confirming the validity of this intrinsic definition of the sidereal zodiac. It is also in conformity with the statement of Cleomedes quoted above who evidently inherited the Babylonian definition.[3] The reconstructed Babylonian star catalogue in Appendix I gives latitudes and longitudes of the 32 Normal Stars together with the latitudes and longitudes of all

---

1. Van der Waerden, private communication after reading an initial draft of this thesis (letter of 30th March, 1983). Cf. also Hunger-Pingree (1999), p 67 for (1) and (2). Here it might seem that van der Waerden is not distinguishing between MUL.MUL (the Pleiades) and Aldebaran. This is not the case. It needs to be borne in mind that Aldebaran, from the Arabic, means 'the follower', i.e. the follower of the Pleiades who lead the way. For the Arabic astronomers, and also for the Indian astronomers, the Pleiades marked the first *manzil* or *nakshatra*, and Aldebaran, with which the moon came into conjunction after the Pleiades, marked the following or second mansion of the moon. Thus the Pleiades and Aldebaran were connected with one another both in Arabic and Indian astronomy, and it can be assumed, as van der Waerden did, that this was also the case in Babylonian astronomy. For the Babylonians, therefore, the rising of MUL.MUL necessarily implied the immediate subsequent rising of Aldebaran.

2. Huber (1958).
3. Neugebauer, HAMA II, p 960.

identifiable stars in Ptolemy's catalogue. The longitudes listed in this reconstruction, defined such that Aldebaran is located at 15° Taurus and Antares at 15° Scorpio, are sidereal, i.e. independent of the location of the vernal point. From this modern reconstruction of the Babylonian sidereal zodiac the relationship between Normal Stars and the zodiac is apparent, e.g. Spica at 29° Virgo marks the closing degree of the sign of Virgo, etc.[1]

HISTORICAL REMARKS

Scientific astronomy originated with the Babylonians. In its observational phase, beginning in the middle of the eighth century BC, dated astronomical observations were recorded. Around 687 BC two astronomical tablets named MUL.APIN were inscribed, each containing a wealth of astronomical observations and also indicating early Babylonian astronomical conceptions such as the three paths (Enlil, Anu, Ea) representing the sun's movement in declination during the course of the year (Figure 11). As discussed in Chapter 3, this schematic solar year indicated by the paths of Enlil, Anu, and Ea was a forerunner both of the Babylonian sidereal zodiac and of Euctemon's tropical calendar—and thus also of the tropical zodiac introduced into astronomy by Hipparchus. Tablet I of MUL.APIN also includes a list of 17 constellations along the path of the moon, referring to the moon 'touching' the stars belonging to these constellations.[2] Since the moon never moves more than 5¼° north or 5¼° south of the ecliptic, all of these 17 constellations contain stars that are within approximately 5¼° of the ecliptic.[3] All of the twelve zodiacal constellations are

---

1. Britton-Walker (1994), p 49, came to the conclusion that "This sidereal zodiac appears to have been fixed so that the longitude of the bright star β Geminorum was 90°. Consequently, the equinoxes and solstices occurred at about 10° of their respective signs in 500 BC." From Appendix I the longitude of Pollux (β Geminorum) in the reconstructed Babylonian sidereal zodiac was 28½° Gemini and not 30° Gemini as suggested by Britton and Walker. However, their conclusion that the vernal equinox was located at 10° Aries in 500 BC exactly fits the reconstructed sidereal zodiac in Appendix I in which Pollux (β Geminorum) is located at 28½° Gemini. Nevertheless, a location of Pollux close to the position suggested by Britton and Walker is attested to in India, where Pollux is located at 29½° Gemini in the Hindu zodiac. This 1° difference is occasioned by a different *intrinsic definition* of the Hindu zodiac from that which evidently was the defining principle of the Babylonian sidereal zodiac. Instead of Aldebaran (15° Taurus) and Antares (15° Scorpio), defining the fiducial axis of the Babylonian sidereal zodiac, in the Hindu zodiac the star Spica is taken as the primary determinant (fiducial star) and is specified to be located at 30° Virgo. This gives rise to a 1° difference with the Babylonian zodiac in which Spica is located at 29° Virgo (see Chapter 6).

2. List VI: MUL.APIN I, IV 31–39. Cf. Hunger-Pingree (1989), pp 62–64 and (1999), p 71.

3. Whereas some of the stars belonging to these 17 constellations lie within 5¼° north or south of the ecliptic, Aldebaran, the main star marking the Babylonian constellation of the 'Bull

named on this list, which can thus be seen as a forerunner of the Babylonian sidereal zodiac. As discussed in Chapter 6, these 17 constellations along the path of the moon were possibly the forerunners of the 28 *nakshatras* (lunar mansions) central to Indian astronomy.

Not long after the MUL.APIN tablets were inscribed, from the middle of the seventh century BC onwards, 'Astronomical Diaries' of recorded observations were kept. The coordinate system used in the Astronomical Diaries—in the early phase, prior to the use of the zodiac as a coordinate system—was a set of 32 Normal Stars, i.e. the brighter stars lying close to the ecliptic.[1] The moon and planets were recorded as being 'above' or 'below' a Normal Star, with distances measured in cubits and fingers.[2] Graßhoff concludes that "above" and 'below' are indicators of latitude (in relation to the ecliptic).[3] Whether a planet lay east or west of the star was also recorded: "The expression 'in front of' a Normal Star means that the planet is located to the west of the star."[4] Graßhoff's conclusion that "topographical relations in the Babylonian Diaries are accurately measured ecliptic coordinates"[5] signifies that not only are 'above' and 'below' indicators of ecliptic latitude, but also 'in front of' and 'behind' specify ecliptic longitude in the directions 'west' and 'east' of the given reference star (Normal Star).

During the fifth century BC a new coordinate system was introduced: the zodiac with twelve 30° signs, which effectively superseded the system of Normal Stars. The zodiac became the basic frame of reference in the later phase of Babylonian astronomy, although the earlier system of Normal Stars continued to be used until the end.[6] Whereas Normal Stars were used solely for observational purposes, the introduction of the zodiac permitted the observations to be treated mathematically.

---

of Heaven', lies 5½° south of the ecliptic and is therefore just outside of the range of the moon's possibility to 'touch' it. Further, two of these 17 constellations, Orion and Perseus, must have been larger for the Babylonians than they are now, since they evidently contained stars belonging to our constellation of Taurus; cf. Hunger-Pingree (1999), p71.

1. Hunger-Pingree (1999), pp148–151 indicate that the Normal Stars were all within 10° north and 7½° south of the ecliptic.
2. 1 cubit = 30 fingers = 2½° is the older norm attested in Babylonian astronomy during the Persian period and before, whereas 1 cubit = 24 fingers = 2° is the norm that was in use during the Seleucid Era; cf. Neugebauer, HAMA II, p591. Note that 1 finger = 0°05' remained a constant unit throughout Babylonian astronomy.
3. Graßhoff (1999), p144.
4. Neugebauer, HAMA I, p546.
5. Graßhoff (1999), p144.
6. Neugebauer, ACT I, p10, n44: "The latest date known from cuneiform texts is an 'almanac' for AD 75."

The advantage of the zodiacal coordinate system was that it provided a continuous scale of measurement and therefore allowed a mathematical treatment of observations, which an isolated set of points, or Normal Stars, did not permit. Moreover, for observational purposes the zodiac is just as effective—if not more so—than the system of Normal Stars. The basic zodiacal unit, the degree, is a natural one since the apparent diameters of the moon and of the sun are approximately half a degree. Hence if a planet is observed to be four "moon's breadths in front of" a Normal Star, the planet is roughly 2° west of the star.

In order for the zodiac to be implemented as a coordinate system the zodiacal longitudes of Normal Stars had to be defined. The remains of a fragment of a Babylonian star catalogue show that this definition took place during the Achaemenid period.[1] From this fragment, together with a comparison of planetary positions recorded simultaneously in relation to Normal Stars and in the zodiac, the longitudes of Normal Stars can be reconstructed. How and by whom the Normal Stars were defined in relation to the zodiac is not known. It was, however, the crucial step in the transition from observational to mathematical astronomy. Because of this step it became possible to translate observational material recorded in the Normal Star system into zodiacal longitudes. Thereby the observational material collected by the Babylonians over many centuries was rendered amenable to analysis. Finally, by means no longer known, the Babylonian astronomers analyzed planetary motions within the coordinate system of the zodiac and arrived at linear arithmetical schemes for predicting the movements of the planets. A complete system of such schemes is evident during the last three centuries BC, coinciding by and large with the Seleucid era in Babylon.

The existence of the mathematical astronomy of the Babylonians remained hidden until the end of the nineteenth century when, through decipherment of astronomical cuneiform texts, it was rediscovered.[2] Increasingly, research into Babylonian mathematical astronomy has revealed that it played an extremely influential role in the history of astronomy. The transmission of the Babylonian sidereal zodiac provides a remarkable example of how Babylonian influence spread.

The ecliptic coordinate system of the sidereal zodiac with twelve signs was first transmitted to the Greeks. Allegedly, Cleostratus was the first Greek to make known the signs of the zodiac around 500 BC.[3] Columella reports that Meton (432 BC) used a sidereal zodiac with the vernal point located at 8°

---

1. Sachs (1952, 2).
2. Initially by Strassmaier and Epping, cf. Neugebauer, HAMA I, pp 348 ff.
3. Pliny, NH II, 31 (trsl. Rackham I, pp 188–189). 500 BC seems too early considering that the zodiac first appeared in Babylonian astronomy "between −463 and −453"; Britton (1999), p 244.

Aries.[1] The reliability of these literary sources is questionable, however.[2] Documentary evidence from Greek sources referring to the zodiac with twelve signs occurs around 300 BC in the writings on spherical astronomy by Autolycus and Euclid.[3] In these writings the zodiac and ecliptic are treated as well-known concepts—without any indication, however, whether they are referring to the sidereal zodiac (Babylonian) or to the tropical zodiac (Greek). At some time during the early Hellenistic period the zodiac was transmitted to Egypt. A demotic ostracon,[4] dated around 250 BC,[5] gives a list of the five 'living stars', i.e. the five planets, and the twelve zodiacal signs. The earliest archeological evidence of the zodiac in Egypt is provided by the Esna zodiac, dating from around 200 BC.[6] It was in Egypt, primarily in Alexandria, that astrology developed.[7] There the influx of Babylonian astronomical methods combined with astral mythology and ideas of predestination gave birth to the complex system of astrology.[8] Although the first known horoscopes were Babylonian,[9] astrological doctrines as they were transmitted throughout the ancient world, were essentially a product of the Hellenistic mind.[10] From Alexandria astrology—as the practice of casting horoscopes for a person's birth—spread far and wide and was transmitted to India at some time in the second century AD.[11] The names of the zodiacal signs in Sanskrit are direct translations of the Greek names and have the same iconography. Indian astronomers did not adopt the re-definition of the zodiac (from sidereal to tropical) introduced by Hipparchus. To the present day the zodiac of Hindu astronomers and astrologers is sidereal,[12] as discussed in Chapter 6.

In the West it is the tropical zodiac introduced into astronomy by Hipparchus that is still employed by astronomers, and which is also used by astrologers since Ptolemy introduced it into astrology in the *Tetrabiblos*. In India,

---

1. Columella, *De re rustica* IX, 14 (trsl. Ash-Forster-Heffner II, pp 487–489). 432 BC would be a possible, albeit early date for the Greeks to know about the zodiac (see previous footnote).

2. Neugebauer, HAMA II, p 593 discusses the unreliability of Pliny's statement, and p 596 refers to the lack of reliability of Columella's report.

3. Autolycus, *On a Moving Sphere* and *On Risings and Settings*; Euclid, *Phaenomena*.

4. Spiegelberg (1902), cols. 6–9. An ostracon was a potsherd or tile used for writing on in ancient Egypt. Concerning demotic, cf. above (p 86, n 2).

5. Neugebauer (1943), pp 121 f.

6. Cf. Neugebauer-Parker, EAT III, pp 72–74.

7. Neugebauer, HAMA I, p 5: "The real center of ancient astrology, from which it eventually spread over the whole world, is undoubtedly Hellenistic Alexandria."

8. Van der Waerden, SA II, pp 127 ff. gives an account of the development of astrology.

9. Rochberg (1998); cf. also Sachs (1952, 1).

10. Bouché-Leclerq, *L'astrologie grecque* gives a comprehensive survey of Greek astrological doctrines.

11. Pingree (1976), p 112; cf. also Pingree (1978).

12. Pillai, *An Indian Ephemeris* I, p 10.

however, the use of the sidereal zodiac, although its origin was lost in obscurity until recently, is a tribute to that innovation of the Babylonians which marked the beginning of mathematical astronomy. For through the innovation of the ecliptic coordinate system of the sidereal zodiac by the Babylonians, the accurate measurement of planetary motions and the derivation of mathematical schemes to describe these motions was made possible. This initial development of mathematical astronomy was thus on account of the zodiac whose *intrinsic definition* is evidently given by the positioning of the stars Aldebaran and Antares at the middle of their respective signs, Taurus and Scorpio. It is this intrinsic definition of the Babylonian zodiac from the early fifth century BC, although it is not known who originated it, which is the central theme of this thesis on the definition and transmission of the zodiac.

6

THE SIDEREAL ZODIAC IN INDIA

THE *NAKSHATRAS* (LUNAR MANSIONS)

[P]ractically all fundamental concepts and methods of ancient astronomy, for the better or the worse, can be traced back either to Babylonian or to Greek astronomy. In other words, none of the other civilizations of antiquity, which have otherwise contributed so much to the material and artistic culture of the world, have ever reached an independent level of scientific thought. Only into astrology were incorporated two remnants of pre-scientific astronomical lore from other than Mesopotamian or Greek background: the 36 Egyptian 'Decans' and the 28 Indian 'Lunar Mansions' (*nakshatras*).[1]

AFTER THE INTRODUCTION OF THE ZODIAC from Mesopotamia into Egypt early in the Hellenistic period, the 36 Egyptian decans became assimilated into the zodiac, three decans to a zodiacal sign, each decan corresponding to 10° in a given sign, as discussed in Chapter 4 concerning the history of the zodiac in Egypt.[2] The Egyptian decans are of considerable antiquity, first appearing on coffin lids around 2150 BC.[3] Similarly, the 28 Indian lunar mansions have a history dating back to around the beginning of the last millennium before Christ. The 28 *nakshatras*, as the Indian lunar mansions are called, comprise a lunar zodiac marking the passage of the moon around the night sky. The *nakshatras* are alluded to in the Vedas:[4]

> Seeking favour of the twenty-eight-fold wondrous ones, shining in the sky together, ever-moving, hasting in the creation, I worship with songs the days, the firmament.[5]

---

1. Neugebauer, HAMA I, p6.
2. Neugebauer-Parker, EAT I, pp95–96, and III, p168.
3. Neugebauer-Parker, EAT I, pp4 ff.
4. There is some controversy over the dating of the Vedas, but it is possible that their date of composition lies anterior to 800 BC.
5. *Arthava-veda Samhita* XIX.7.1 (trsl. Whitney, p906).

Figure 17

The 28 Nakshatras

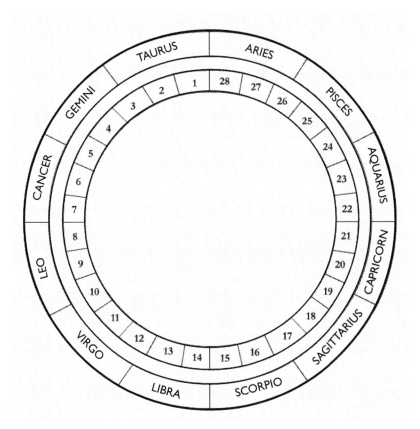

An exact definition of the lunar zodiac of *nakshatras* in relation to the Babylonian zodiac is not known, and therefore Figure 17 represents a schematic relationship of the 28 *nakshatras* in connection with the twelve signs of the Babylonian sidereal zodiac. In the Vedas the 28 *nakshatras* are not defined precisely in relation to the fixed stars, nor are the *nakshatras* necessarily of equal length. In order that the stellar determinants of each *nakshatra* may be readily found, however, the 28 *nakshatras* are assumed to be of equal length, each 12°51'26" long, with the 1st *nakshatra*, Krittika, beginning at 0° Taurus in the Babylonian zodiac. The zero point of the lunar zodiac is here defined to coincide with 0° Taurus not only for the sake of convenience but also because this choice yields a close agreement with the known stellar determinants of *nakshatras* (Table 6). Using Figure 17 in conjunction with the reconstructed Babylonian star catalogue (Appendix 1), the fixed stars belonging to each *nakshatra* are easily found.

As in the Egyptian decan lists, the earliest *nakshatra* lists associate each lunar mansion with a presiding deity (Table 6), e.g. the deity of Krittika, the first *nakshatra* in the Vedic lists, is Agni.[1] The Pleiades are the stellar determinants of Krittika.[2] This fact alone, that the Pleiades are associated with the beginning of the lunar zodiac (of 28 *nakshatras*),[3] suffices to distinguish it from the Babylonian zodiac, whose premier sign is marked by the stars of Aries. Interestingly, the Babylonian astronomers did draw an association between the Pleiades and the moon, ascribing the 'exaltation' of the moon to this star cluster.[4] Here it is reasonable to suppose that Babylonian astronomy exerted an influence upon

---

1. Dumont (1954), p 205. Cf. also Hunger-Pingree (1999), p 63, where Pingree remarks that, "The Mesopotamian association of gods with constellations in the late second millennium BC probably gave the idea to the Vedic Indians to associate one or a set of their gods to each of the twenty-eight *nakshatras* or constellations alleged to be in the path of the Moon."

2. *Surya-Siddhanta* (trsl. Burgess, p 324).

3. After the introduction of the zodiac into India, probably in the second century AD, the *nakshatras* became redefined. Asvini, with marking stars β and γ Arietis, became the premier *nakshatra*, and one of the *nakshatras* became discarded, leaving 27 *nakshatras* in the later Indian astronomy from around the second century AD onwards.

4. Weidner (1919) discusses the Babylonian *qaqqar nisirti*, which are precisely the *hypsomata* or exaltations of Greek astrology. Kugler, SSB I, p 40 refers to a Babylonian text listing the *qaqqar nisirti* of the planets. In the Babylonian text each planet's *hypsoma* is in a sign of the zodiac, e.g. for Jupiter it is in the sign of the Crab, or between the sign of the Crab and the sign of the Lion. In the three *hypsomata* depictions published by Weidner (1919), plate v, the engraved drawings each depict a planet in relation to a sign of the zodiac and to stars within that sign, e.g. the moon is drawn in Taurus adjacent to a cluster of seven stars. The identification of these seven stars as the Pleiades cluster is certain, since the name 'MUL.MUL' is engraved on the drawing and from other texts it is known that MUL.MUL means the Pleiades. Cf. also Hunger-Pingree (1999), p 28, who transcribe *asar nisirti* or *bīt nisirti* for hypsoma (exaltation) instead of *qaqqar nisirti*. Concerning exaltations Pingree remarks that "The earliest attestation of this concept is found in the inscriptions of Esarhaddon of Assyria" (–679 to –667).

The signs of exaltation are well known in Greek astrology. They are referred to by Vettius Valens, Firmicus Maternus, Dorotheus Sidonius and Sextus Empiricus, cf. Robbins (1936), p 106 and Bouché-Leclerq, *L'astrologie grecque*, pp 184–198. The Greek tradition of *hypsomata* refers to specific positions of the planetary exaltations, generally given as (cf. Mackenzie [1964], p 524):

Sun: ♈ 19°   Saturn: ♎ 21°   Mars: ♑ 28°   Mercury: ♍ 15°
Moon: ♉ 3°   Jupiter: ♋ 15°   Venus: ♓ 27°

Moreover, exactly the same positions are recorded by Muslim astrologers, who add to the list exaltations for the lunar nodes (cf. Pingree, *The Thousands of Abū Ma'shar*, p 61):

Ascending Node ♊ 3° Geminorum
Descending Node ♐ 3° Sagittarii

From the reconstruction of the Babylonian star catalogue (Appendix 1) it is evident that the moon's exaltation position, 3° Taurus, is adjacent to the Pleiades cluster, exactly as depicted in the Babylonian *hypsoma* engraving.

Indian astronomy, whereby the Babylonian association of the Moon with the Pleiades influenced Indian astronomers in their choice of the Pleiades *nakshatra* (Krittika) as the first lunar mansion. In an early *nakshatra* list, from the *Taittiriya-Brahmana*, the first 14 *nakshatras* beginning with Krittika are called 'Deva' *nakshatras* and the remaining *nakshatras* (15–28) are called the 'Yama' *nakshatras*.[1] Accordingly, Krittika (Pleiades) heads the list and separates the Deva from the Yama *nakshatras*.

The moon passes through the zodiac of 28 *nakshatras* in one (sidereal) month of ca. 27⅓ days, spending approximately 24 hours in each *nakshatra*. In the course of this '*nakshatra* month' the moon becomes full in a particular *nakshatra*. Generally the lunar synodic months in ancient Hindu culture were named after the *nakshatra* (or its neighbor) in which the moon became full, e.g. according to tradition Gautama Buddha was born at the full moon in the (lunar) month of Vaisakka,[2] i.e. when the full moon stood in the 14th *nakshatra*, Visakha, whose stellar determinants are the 'balance pans' α and β Librae (Figure 17 and Table 6).

The use of lunar asterisms in astronomy and astrology is evident not only in India but also in China,[3] Persia,[4] Greece,[5] and Egypt.[6] This has led to discussion as to whether the *nakshatras* might have originated elsewhere, possibly in Mesopotamia.[7] Fairly conclusive evidence that this is indeed the case is presented by Pingree, who points out the similarity of the Indian *nakshatra* lists with a star list (List VI) of 17 constellations in the path of the moon from the Babylonian MUL.APIN tablet I, already referred to in Chapter 5 as a forerunner of the Babylonian sidereal zodiac:

> The Indian lists of *nakṣatras* composed during the early centuries of the last millennium BC show striking resemblances to the Mesopotamian constellations, and to MUL.APIN's List VI in particular, though the Indians, wishing to have one *nakṣatra* for the moon to spend each night of a sidereal month with, have split some Mesopotamian constellations in two and have added a number of *nakṣatras* with very high latitudes, far beyond the observed path of the moon (see Pingree [1989], pp439–442). The following list of Indian *nakṣatras* with their *yogatārās* (junction stars) as of the fifth century AD will illustrate the

---

1. Dumont (1954), p205.
2. Pillai, *An Indian Ephemeris*, p477.
3. Needham, *Science and Civilization in China* III, pp252–259.
4. Albiruni, *The Chronology of Ancient Nations* (trsl. Sachau, pp226–228).
5. Weinstock (1949).
6. Ibid., pp51f.
7. Kane, *History of Dharmasāstra* v, 1, pp508 f., discusses whether the *nakshatras* are indigenous to India or borrowed from another culture.

relationship more clearly (the identifications of the *yogatārās* are those of Pingree-Morrissey [1989])

	Nakṣatra	*Yogatārā*	*Babylonian constellation*
1	Kṛttikā	η Tauri	Stars
2	Rohiṇī	α Tauri	Bull of Heaven
3	Mṛgaśiras	λ Orionis	True Shepherd of Anu
4	Ārdrā	α Orionis	True Shepherd of Anu
5	Punarvasu	β Geminorum	Great Twins
6	Puṣya	δ Cancri	Crab
7	Āśleṣā	κ Cancri(?)	Crab
8	Maghā	α Leonis	King
9	Pūrva Phalgunī	δ Leonis	Lion
10	Uttara Phalgunī	β Leonis	Lion
11	Hasta	δ Corvi	Raven
12	Citrā	α Virginis	Furrow
13	Svāti	α Boötis	ŠU.PA
14	Viśākhā	ι Librae	Scales
15	Anurādhā	δ Scorpii	Scorpion
16	Jyeṣṭhā	α Scorpii	Scorpion
17	Mūla	45 Ophiuci	Ophiuchus
18	Pūrva Aṣāḍhā	δ Sagittarii	Pabilsag
19	Uttara Aṣāḍhā	ζ Sagittarii	Pabilsag
20	Abhijit	α Lyrae	She-goat
21	Śravaṇa	α Aquilae	Eagle
22	Dhaniṣṭhā	α Delphini	Delphinus
23	Śatabhiṣaj	λ Aquarii	Great One
24	Pūrva Bhadrapadā	α Pegasi	Field
25	Uttara Bhadrapadā	α Andromedae	Field
26	Revatī	ζ Piscium	Anunītu
27	Aśvinī	β Arietis	Hired Man
28	Bharaṇī	35 Arietis	Hired Man

Hunger-Pingree (1999), p 72.

As noted by Pingree, the junction stars (*yogatārās*) of *nakshatras* 11, 13, 17, 20, 21, 22, 24, and 25 all lie beyond the path of the moon, so these *nakshatras* are additional to the Mesopotamian list of 17 constellations in the path of the moon. Moreover, in the *nakshatra* list there are three instances in which a single constellation is divided into two *nakshatras* denoted by *pūrva* (prior) and *uttara* (posterior). Possibly in this way, from the 17 Mesopotamian constellations in the path of the moon, 28 constellations (*nakshatras*) in the path of the moon were derived. This tentative finding by Pingree raises the question concerning the degree of cross-cultural transmission between Mesopotamia and

India during the first millennium BC by means of which the 17 Mesopotamian lunar constellations might have been transmitted to India and then adapted there as the system of 28 *nakshatras*. That there was cross-cultural transmission during the time of the Achaemenid occupation of the Indus valley between approximately 513 and 326 BC is known from the appearance of Mesopotamian astronomical methods and parameters in Lagadha's *Jyotisavedanga* from about the fifth century BC.[1] Therefore, unless the *nakshatra* system developed in Indian astronomy independently of the Mesopotamian constellations, it now seems possible that the *nakshatra* system developed in India under the influence of knowledge of the 17 Mesopotamian constellations in the path of the moon listed in MUL.APIN.[2] In this case, for the history of the zodiac it would signify that the Mesopotamian constellations observed in relation to the moon—the forerunners of the twelve signs of the zodiac—later, in the guise of the 28 Indian *nakshatras*, were integrated into the Babylonian zodiac, but then only as 27 *nakshatras*, when the zodiac was transmitted to India around the second century AD, as discussed below.

The use of lunar asterisms, named *manzils*, by Islamic astronomers is evidence of the Indian influence in Islamic astronomy. The emergence and development of Islamic astronomy has a complicated background. In addition to the influence of Indian astronomy, there is no doubt that the Islamic astronomers were recipients of Persian astronomical and astrological literature stemming from the time of the Sasanian kings.[3] Islamic science also incorporated the Ptolemaic system, received through Greek sources, and in the ninth century the *Almagest* became translated into Arabic.[4] However, the primary impulse towards the development of Islamic astronomy appears to have been from India. Hindu astronomical treatises, e.g. Brahmagupta's *Khandakhadyaka* written in AD 665, known as the *al-Arkand* to the Muslims, played an important part in this development.[5] Arabic verses in which weather lore is related to the full moon[6] in a given *manzil* reflect a similar usage of the *manzils* by the Muslims to that of the *nakshatras* by the Hindus, although in Indian culture it was especially the ceremonial aspect of the *nakshatras*, with each *nakshatra* being sacred to some deity, that was important (Table 6).

---

1. Pingree (1973).
2. Pingree (1989).
3. Neugebauer, HAMA I, p8.
4. Ibid.
5. Ibid., p7.
6. Albiruni, *The Chronology of Ancient Nations* (trsl. Sachau, pp 226–228), e.g., "When the full moon joins Aldabarān in the 14th night of a month, then winter encircles the whole earth."

Table 6

The 28 Nakshatras and Their Deities

	Ecliptic Longitude	Nakshatra	Vedic Deity
1	♉ 00°00'–♉ 12°51'	Krittika	Agni
2	♉ 12°51'–♉ 25°43'	Rohini	Prajapati
3	♉ 25°43'–♊ 08°34'	Mrigasiras	Soma
4	♊ 08°34'–♊ 21°26'	Ardra	Rudra
5	♊ 21°26'–♋ 04°17'	Punarvasu	Aditi
6	♋ 04°17'–♋ 17°09'	Pushya	Brhaspati
7	♋ 17°09'–♌ 00°00'	Aslesha	Sarpah (the Serpents)
8	♌ 00°00'–♌ 12°51'	Magha	Pitarah (the Fathers)
9	♌ 12°51'–♌ 25°43'	Purvaphalguni	Aryaman
10	♌ 25°43'–♍ 08°34'	Uttaraphalguni	Bhaga
11	♍ 08°34'–♍ 21°26'	Hasta	Savitar
12	♍ 21°26'–♎ 04°17'	Chitra	Tvastar*
13	♎ 04°17'–♎ 17°09'	Svati	Vayu
14	♎ 17°09'–♏ 00°00'	Visakha	Indra and Agni
15	♏ 00°00'–♏ 12°51'	Anuradha	Mitra
16	♏ 12°51'–♏ 25°43'	Jyestha	Indra
17	♏ 25°43'–♐ 08°34'	Mula	Nirrti*
18	♐ 08°34'–♐ 21°26'	Purvashadha	Apah (the Waters)
19	♐ 21°26'–♑ 04°17'	Uttarashadha	Vishve-Devah
20	♑ 04°17'–♑ 17°09'	Abhijit	Brahma
21	♑ 17°09'–♒ 00°00'	Sravana	Vishnu
22	♒ 00°00'–♒ 12°51'	Dhanistha	Vasavah (the Vasus)
23	♒ 12°51'–♒ 25°43'	Satabhisaj	Varuna*
24	♒ 25°43'–♓ 08°34'	Purvabhadrapada	Aja Ekapad
25	♓ 08°34'–♓ 21°26'	Uttarabhadrapada	Ahi Budhniya
26	♓ 21°26'–♈ 04°17'	Revati	Pusan
27	♈ 04°17'–♈ 17°09'	Asvini	Asvinau (the 2 Asvins)
28	♈ 17°09'–♉ 00°00'	Bharani	Yama

See Figure 17 for the schematic definition of the lunar *nakshatras* in relation to the Babylonian zodiac as indicated in Table 6. In this definition it is assumed that the *nakshatras* are of equal length, each 12°51'26" long, with the first *nakshatra*, Krittika, beginning at 0° Taurus in the Babylonian sidereal zodiac. This definition conforms well with the stellar determinants of the nakshatras, cf. *Surya-Siddhanta* (trsl. Burgess, p 324) and with the stellar determinants of the corresponding *manzils*, cf. Kunitzsch, *Arabische Sternnamen in Europa*, pp 55–56. The Vedic deities are those given by Kane, *History of Dharmasastra* v, 1, pp 501–504, except those marked with an asterisk (*), which are duplicates in Kane's list and are therefore replaced by the deities listed by Dumont (1954), p 205. Otherwise the two lists are identical, with the exception of certain of Dumont's renderings (given in parentheses).

THE SIDEREAL ZODIAC IN INDIA 121

The original use of *nakshatras* by the Vedic priests was ceremonial, e.g. the text of the *Baudhayans* begins, "I prepare the oblation agreeable to Agni, to the Kṛttikās" (*Baudh.* 28.3).[1] However, with the naming of the lunar months according to the *nakshatra* in which the full moon occurs, the *nakshatras* (together with the moon's phase) became related to the passage of the seasons. Hence seasonal conditions and the weather were referred to the succession of lunar months. In this way the moon's phase in relation to the Indian *nakshatras* or to the *manzils* of Islamic astronomy served as an indicator of the weather and of the seasons.

THE REDEFINITION OF THE *NAKSHATRAS*

The definition of the 28 *nakshatras* in terms of ecliptic longitudes in the Babylonian sidereal zodiac (Table 6) is schematic, yet it corresponds well with how the *nakshatras* were originally defined in Indian astronomy. Before the introduction of the zodiac into India in the second century AD, the Vedic priests referred to the *nakshatras* by means of reference stars (*yogatārās*). They used a similar system to that of the Babylonians. As discussed in Chapter 5, in the first period of Babylonian astronomy, Babylonian astronomers employed a system of reference stars (Normal Stars) before they began to use the zodiac with twelve 30° signs. The zodiac essentially superseded the system of Normal Stars from around the fourth century BC onwards, although the Normal Stars were sometimes used as well, simultaneously with the sidereal zodiac, in referring to planetary positions in signs and degrees of the zodiac. Likewise, the Vedic priests referred to the moon in a given *nakshatra* simply by observing the position of the moon in relation to the reference star(s) of each *nakshatra*.

As already mentioned, the Pleiades are the reference stars for Krittika (*nakshatra* 1), and the Pleiades cluster is located about 4° north of the ecliptic. Some of the reference stars, however, lie very far from the ecliptic, e.g. Vega, the reference star for Abhijit (*nakshatra* 20), is about 62° north of the ecliptic. Thus, whereas the Babylonian longitude of the Pleiades (5° Taurus) lies within the interval for the *nakshatra* Krittika listed in Table 6, that of Vega (20½° Sagittarius) does not fall within the interval indicated for the *nakshatra* Abhijit given in Table 6.[2] Vega is an exception, however, as most of the stellar determinants of the *nakshatras* do fall within the limits as given in Table 6.[3] The difference between Babylonian and early Indian giving rise to differences in

---

1. Dumont (1954), p206.
2. The longitudes given for the Pleiades and for Vega are those of the reconstructed Babylonian sidereal zodiac (Appendix I).
3. *Surya-Siddhanta* (trsl. Burgess, p324).

some stellar locations is due to the use of polar longitudes in Indian astronomy as opposed to ecliptic longitudes in the sidereal zodiac used by the Babylonians from the fifth century BC onwards. For stars far north of the ecliptic, such as Vega, this means that the discrepancy between polar and ecliptic longitudes increases the further north the star is located.[1]

With the introduction into India of the new coordinate system, the zodiac, probably during the second century AD, the *nakshatras* became redefined. A new list was drawn up in which the *nakshatras* became related to the signs of the zodiac.[2] In the process of their assimilation to the zodiac the *nakshatras* acquired a new definition. In the new list only 27 *nakshatras* are enumerated and Asvini is the premier nakshatra, replacing Krittika at the head of the list.

> [The] nakshatra lists begin with the Krittikas in the Vedic literature and ... with Asvini in classical Sanskrit literature....[3]

The difference between the Vedic list of 28 *nakshatras* and the modified list of 27 is that Abhijit, the twentieth nakshatra, is omitted, and the zero point (taken to be the beginning of Asvini) is defined to be 0° Aries. The 27 *nakshatras*, each 13°20' in length, are thus related to the zodiac, with Asvini extending from 0° Aries to 13°20' Aries, Bharani from 13°20' Aries to 26°40' Aries, etc.[4]

The disappearance of Abhijit, the twentieth *nakshatra* in the Vedic list, to yield the modified list of 27 *nakshatras* is explained in mythological terms in *Vanaparva* 230, 2–11.

> Abhijit, the younger sister of Rohini, coveted the position of eldership and went to a forest for practising *tapas*.[5]

In fact, 27 *nakshatras* provides a better approximation to the motion of the moon, since the moon's sidereal period—the length of time taken to return to

---

1. A star's *ecliptic longitude* is determined by the intersection with the ecliptic of the great circle through the star connecting the poles of the ecliptic. For Babylonian astronomers the ecliptic longitude was sidereal, measured in the framework of the Babylonian sidereal zodiac (Appendix I), whereas for Ptolemy in his star catalogue ecliptic longitudes are tropical, measured from the vernal point. In contrast, a star's *polar longitude* is specified by the intersection with the ecliptic of the great circle through the star connecting the celestial poles. For Indian astronomers the polar longitude was sidereal, measured in the framework of the Indian sidereal zodiac, whereas for Hipparchus, who used polar longitudes in his *Commentary to Aratus*, polar longitudes were tropical, measured from the vernal point.
2. This development parallels that of the assimilation of the Egyptian decans into the zodiac discussed in Chapter 4.
3. Kane, *History of Dharmaśāstra* v, 1, p507.
4. Brennand, *Hindu Astronomy*, plates V and VI, pp37–38, show figures of this division, together with reference star longitudes.
5. Kane, *History of Dharmaśāstra* v, 1, p497, n725.

conjunction with the same fixed star—is 27⅓ days. Since the sidereal period of revolution is closer to 27 rather than 28 days, the use of a '*nakshatra* month' of 27 days—in which the moon spends one day in each nakshatra—is closer to a correct astronomical description of the moon's sidereal motion than a *nakshatra* month of 28 days. The use of either 27 or 28 *nakshatras* is an approximation, but 27 is closer to 27⅓ than 28. Nevertheless, the original contribution of the Vedic priests to astronomy was the list of 28 *nakshatras* as a descriptive device for following the passage of the moon through the stars. Even if the *nakshatra* system was an adaptation of the Mesopotamian constellations in the path of the moon (List VI of MUL.APIN), it is still to the credit of the Vedic priests that their *nakshatra* system—or something similar to it—was utilized not only in India (where it continues to be utilized to the present day) but also in other cultures, in particular by Islamic astronomers.

Islamic astronomers took over the system of 28 *nakshatras*, but with the modification of the premier *nakshatra* (*manzil*) as Asvini (Al Sharatan). The Islamic term *manzil* means 'mansion of the moon'. The 28 *manzils* are each 12°51'26" in length.[1] Albiruni (973–1048) collected several lists of the 28 *manzils*—in Arabic, Sogdian, and Chorasmian.[2]

Figure 18

Nakshatras in Indian and Islamic Astronomy
(Historical Summary)

1ˢᵗ millennium BC
28 *nakshatras*
commencing with Krittika
(Vedic priests)

2ⁿᵈ century AD
27 *nakshatras*
commencing with Asvini
(Indian astronomers under
Greco-Babylonian influence)

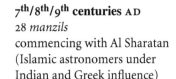

7ᵗʰ/8ᵗʰ/9ᵗʰ centuries AD
28 *manzils*
commencing with Al Sharatan
(Islamic astronomers under
Indian and Greek influence)

1. Kunitzsch, *Arabische Sternnamen in Europa*, pp 55–56 gives a list of identifying stars.
2. Albiruni, *The Chronology of Ancient Nations* (trsl. Sachau, pp 226–228).

The Chorasmian list of 28 lunar mansions commences with Al Thurayya, which corresponds to Krittika, whereas the other lists start with Al Sharatan (corresponding to Asvini). The Chorasmian list seems to indicate that although the *nakshatras* became redefined with Asvini replacing Krittika as the premier *nakshatra*, this redefinition was not universal. The Chorasmian list of *manzils* is an exception, however.

THE INTRODUCTION OF THE ZODIAC INTO INDIA

The development of astronomy in India has an exceedingly complex history that only recently has begun to be uncovered.[1] The evolution of Indian astronomy in antiquity appears to follow four principal stages of development.

The first is that of the Vedic priests, in which the system of *nakshatras* for observing the moon was used.

> [T]he Vedas and Brahmanas provide us with some crude elements of observational astronomy, such as the standard list of 27 or 28 *nakshatras* or constellations associated with the Moon's course through the sky....[2]

The use of the division into *nakshatras* was later extended from being simply a daily record of the moon's location among the stars to an astronomical system for recording the positions of the planets and the lunar nodes, e.g. in the astronomical references from the *Mahabharata*.

> Mars is retrograde in Magha (8th *nakshatra*, marked by Regulus).... The planet Sukra (Venus) rises towards Purva Bhadra (24th *nakshatra*), shining brilliantly and looking towards Uttara Bhadra (25th *nakshatra*). Ketu (lunar descending node), blazing like smoky fire, stops and afflicts the effulgent constellation of Indra (Jyestha, 16th *nakshatra*, marked by Antares).[3]

None of the astronomical citations in the *Mahabharata*, probably composed sometime during the last few centuries BC, refer to the zodiacal signs. Thus the earliest Indian astronomy was based solely on the division into *nakshatras*.

Around the same time the *Mahabharata* was being composed, a new stage in Indian astronomy began. This was the development of a primitive mathematical astronomy "in the fifth or fourth century BC on the basis of information about originally-Mesopotamian methods and parameters transmitted to

---

1. Primarily through the Pingree's research, whose achievement in the field of Indian astronomy parallels that of Neugebauer in Babylonian and Egyptian astronomy; cf. Pingree (1976), (1978), (1989), and Pingree-Morrissey (1989).
2. Pingree (1973).
3. *Bhishma Purva* 3, 14–16 (trsl. Dutt VI, p 4).

India during the Achaemenid occupation of the Indus Valley between ca 513 and 326 BC."[1] As already mentioned, Babylonian-influenced mathematical astronomy is evident in the *Jyotisavedanga* written by Lagadha probably in the fifth century BC.

> [T]he *Jyotisavedanga* [is] one of the six angas or 'limbs' studied by Vedic priests; its purpose was to provide them with a means of computing the times for which the performances of sacrifices are prescribed, primarily new and full moons.[2]

The third level of development of Hindu astronomy began in the second century AD when "the zodiacal signs [were] introduced into India by Greek sources, as is indicated by their Greek names and iconography...."[3] The Sanskrit names for the twelve rasis, or signs of the zodiac (Mesha, Rishabha, Mithuna, Karkataka, etc.), are direct translations of the Greek names (Kriya, Tauro, Jimuthro, Kankri, etc.). It was during the second and third centuries AD that Greek astrological texts began to be translated into Sanskrit, thereby introducing the twelve signs of the zodiac into Indian culture. For example, in AD 149 or 150 Yavanesvara, the 'Lord of the Greeks', translated a Greek astrological text written in Egypt—probably in Alexandria—which dealt with a Greek adaptation of the Babylonian astronomical method for computing the rising-times of the zodiacal signs.[4] (This is an important element in astrology known as the 'horoscope', i.e. the sign of the zodiac rising on the eastern horizon at the moment of birth.)

Thus a new impetus for the development of Indian astronomy came through the spread of astrology which, although originating in Mesopotamia, was developed into a comprehensive system by the Greeks in Hellenistic Egypt.

The spread of astrology depended upon the acquisition of certain basic astronomical concepts required for the computation of the horoscope and the positions of the planets in the various signs of the zodiac. For this a higher level of mathematical astronomy was required than that of the *Jyotisavedanga*. The more advanced mathematical astronomy that became introduced into India through the spread of astrology may be termed 'Greco-Babylonian' in contrast to the Babylonian-influenced astronomy of the *Jyotisavedanga*.

1. Pingree (1973), p 3.
2. Ibid., p 1.
3. Pingree (1976), p 112.
4. Ibid., pp 110–112.

LATER INDIAN ASTRONOMY

The fourth stage of development, superseding the Hindu adaptation of Greco-Babylonian astronomy, was the 'Greek' period of Indian astronomy beginning in the fifth century AD.

> This Greek period begins in the fifth century AD, and is characterized as far as planetary theory is concerned by the use of the *yuga* system (inspired by the Greek *anni magni* and realized in the Keskinto Inscription) for determining mean longitudes, and by the use of geometrical models to explain the planets' deviations from those mean longitudes.[1]

The *yuga* system of the Indians was an altogether remarkable development and continues to be used even now,[2] although the impact of modern western astronomy has begun to undermine the Hindu astronomical tradition. Of the various *siddhantas* ('astronomical treatises') expounding the *yuga* system, the (new) *Surya siddhanta* is the most widely used.[3] The *Surya siddhanta* is said to have been written under the divine inspiration of the Sun-god (*Surya* = Sun); and various other *siddhantas* are also attributed with a divine origin. One exposition of the *yuga* system belongs to "a Brahmapaksa text, the *Paitama-hasiddhanta*, [which] existed in the early fifth century...."[4] It uses a Great Year of 4,320,000,000 years. This was followed by the *Aryabhatiya* of Aryabhata (epoch AD 499), which uses a Great Year of 4,320,000 years. The latter period is known as a *mahayuga* whilst the former period, i.e. 1000 *mahayugas*, is a *kalpa*. The ninth century Islamic astrologer Abū Ma'shar employed a Great Year of 360,000 years, which is one-twelfth of a *mahayuga*.[5]

How the *yuga* system originated is unclear. However, the idea of a Great Year after which all the planets return to their original positions and a new cycle begins is found in Plato's *Timaeus*, and may be of Pythagorean or even Babylonian origin.[6]

> The perfect number of time is accomplished and the perfect year is complete when all the eight revolutions [of all the stars] return to their point of departure.[7]

---

1. Pingree (1976), p115.
2. Pillai, *An Indian Ephemeris*.
3. *Surya-Siddhanta* (trsl. Burgess).
4. Pingree (1976), p115.
5. Kennedy-van der Waerden (1963) discuss the relationship between the Great Year of the Persians and those of the Indians (*mahayuga* and *kalpa*); cf. also Pingree, *The Thousands of Abū Ma'shar*, esp. pp28–30, for a summary of the Persian and Indian systems.
6. Van der Waerden (1952).
7. Duhem, *Le système du monde* I, p67, concerning the great year implied by Plato's exposition in the *Timaeus*.

The eight revolutions are those of Moon, Mercury, Venus, Sun, Mars, Jupiter, Saturn, and the eighth sphere, the sphere of the fixed stars. Plato does not say what the perfect number is, but indicates a formula for its calculation which is, however, ambiguous and unclear.[1] Different values for the Great Year exist, e.g. the world-year of the Persians, 360,000 years according to Abū Ma'shar, differs from that of the Indians, which is either a *mahayuga* of 4,320,000 years or a *kalpa* (=1000 *mahayuga*) of 4,320,000,000 years. Nevertheless, there is general agreement in the Indian and Persian sources upon a fundamental principle concerning the definition of the Great Year:

> A world-year, according to the generality of the astrologers, is from the time of arrival of the planets at the first of Aries until the time of their return to the end of Pisces....[2]

If the number of revolutions that each planet makes around the zodiac during a world-year is known, then the positions of the planets at any point in time can be determined. The number of revolutions completed since the beginning of the world-year is first calculated and then a residual component for any part of a cycle not yet completed is added. This principle is utilized in the *kalpa* system, the *mahayuga* system, and the system of Abū Ma'shar.

> Abū Ma'shar's astronomical system was based on the assumption of a series of mean conjunctions of the planets at 0° Aries (Grand Conjunctions) spaced at equal intervals in time. The mean longitudes of the planets for any given date can be determined once it is known how many revolutions each mean planet makes between successive Grand Conjunctions and how many revolutions each mean planet makes between successive Grand Conjunctions and how much time has elapsed since the last Grand Conjunction.[3]

In the *kalpa* system not only the mean planets but also their nodes and apogees are assumed to be at 0° Aries at the beginning of the world-year, where the nodes and apogees are endowed with a slow motion around the zodiac in the allotted period. Two variants of the *mahayuga* system are to be distinguished. Both are associated with Aryabhata (b. 476),[4] and in both it is the mean planets only (and not their nodes and apogees) that are at 0° Aries initially. The two *mahayuga* systems are known as the 'midnight' and 'sunrise' systems, respectively. In the former a Grand Conjunction of the mean planets is assumed to have taken place at midnight of Thursday/Friday, 17/18 February

---

1. Ibid., p84.
2. Kennedy-van der Waerden (1963), p316.
3. Pingree, *The Thousands of Abū Ma'shar*, pp27–28.
4. Pingree, *Census of the Exact Sciences in Sanskrit* A, I, 50B–53B and II, 15B, and *Dictionary of Scientific Biography* I, pp308–309.

–3101. In the sunrise system the Grand Conjunction is dated to sunrise on Friday, 18 February –3101. The difference between the two systems depends upon whether the day is taken to start at midnight or sunrise. Thursday, 17 February –3101 is, according to the tradition cited by Aryabhata, the Thursday of the 'battle of Bharata', the great war between the Kurus and the Panchalas described in the *Mahabharata*.

The new era, *Kaliyuga*, began on the next day, i.e. Friday. According to whether the day is conceived of as beginning at midnight or sunrise, the start of *Kaliyuga* is reckoned to have begun either at midnight from Thursday to Friday, or at sunrise on Friday, 18 February –3101.[1]

The method of Indian astronomers employing the *yuga* system (e.g. Aryabhata) is conceptually straightforward and simple to apply. Starting with the Grand Conjunction of the planets at 0° Aries (sidereal) at the beginning of *Kaliyuga*, the cycles of the planets in their revolutions of the zodiac are followed, so that at any given point in time, measured from the start of *Kaliyuga*, the number of complete cycles of each planet can be determined and then the mean position of each planet in its current cycle, expressed as a fraction of the complete cycle. Corrections are then applied to the mean positions to find the actual positions of the planets in the zodiac.[2]

At the end of a Great Year all the planets will have revolved through a certain number of complete cycles and returned to their starting point at 0° Aries in the sidereal zodiac. If the number of zodiacal revolutions of each planet in a Great Year is known, then the mean sidereal period of each planet can be calculated, and the number of complete cycles in a given time is then readily computed.

THE INDIAN ASTRONOMICAL
TRADITION OF THE GREAT YEAR

There is an important difference in conception between the Great Year of the Platonic tradition[3] and the Great Year of the Indian astronomers.[4] Plato

---

1. This is not the only difference between the midnight and sunrise systems. The *mahayuga* of the midnight system comprises four unequal *yugas* whose ratios to each other are 4:3, 3:2, and 2:1 (*Kritayuga* = 1,728,000; *Tretāyuga* = 1,296,000; *Dvāparayuga* = 864,000; *Kaliyuga* = 432,000 years) whilst the lengths of the *yugas* in the sunrise system are all equal to 1,080,000 years. Cf. Pingree, *The Thousands of Abū Ma'shar*, pp 27–28.

2. Billard, *L'astronomie indienne*, pp 113–114.

3. Duhem, *Le système du monde* I, pp 65–85, pp 284–296; II, pp 214–223, gives an account of this tradition.

4. Kennedy (1964) outlines the ramifications of the Indian world-year concept in Islamic astrology.

specifies the completion of the planetary revolutions and the sphere of the fixed stars, the eighth sphere. Since Plato lived before Hipparchus, the conception of the precession of the equinoxes was unknown to him. However, the expression 'revolution of the eighth sphere' subsequently became identified with the phenomenon of precession, since within the framework of the tropical zodiac of the Greek tradition the fixed stars are usually considered to revolve in relation to the equinoxes. Thus the Platonic Great Year came to be thought of in relation to the precession of the equinoxes or, rather, to the revolution of the eighth sphere.[1]

There is a fundamental difference, however, between the conception 'precession of the equinoxes' and the conception 'revolution of the eighth sphere'. This difference highlights the dichotomy between Indian astronomy—which is sidereal—and later Greek (and western) astronomy, which is tropical. If the sidereal zodiac, e.g. that of the Babylonians, is used astronomically as the primary coordinate system, then the conception of precession is natural, i.e. the steady shift of the vernal point retrogressively through the signs of the zodiac is evident. If, however, the tropical zodiac introduced into astronomy by Hipparchus is taken to be the primary coordinate system, then it is evident that "the sphere of the fixed stars performs a rearward [eastward] motion along the ecliptic."[2] Thus, as discussed in Chapter 2, Hipparchus observed that the star Spica had shifted 2° eastward in relation to the autumnal point between Timocharis and his own time (about 150 years). Since Hipparchus referred these measurements to the equinoctial points, he observed that the sphere of the fixed stars had rotated through 2°. However, if a Babylonian astronomer had made the same observation, he might have expressed that at the time of Timocharis the vernal point was located at 7° Aries and at the time of Hipparchus at 5° Aries,[3] i.e. instead of observing that the sphere of the fixed stars had rotated through 2°, he would say that the vernal point had retrogressed 2° in the zodiac.

Hipparchus himself was apparently in two minds as to how to view the phenomenon that he had discovered. On the one hand—according to Ptolemy—he maintained that:

---

1. The Neoplatonist Proclus states that the great year is determined by the length of time taken for the planets (including the sun and moon) to return to conjunction at the same equinoctial or tropical point; cf. Duhem, *Le système du monde*, p 292.
2. The title of chap. 2, bk 7, in the *Almagest*: Ptolemy, *Almagest* VII, 2 (trsl. Toomer, p 327).
3. Assuming that Spica's sidereal longitude is 29° Virgo (Appendix I), then Timocharis' observation of Spica 8° west of the autumnal point located the autumnal point at 7° Libra and thus the vernal point at 7° Aries in his day. Similarly, Hipparchus' observation located the vernal point at his time at 5° Aries.

Spica was formerly 8°, in zodiacal longitude, in advance of the autumnal [equinoctial] point, but is now 6° in advance.[1]

On the other hand (again according to Ptolemy):

> In the work *On the Length of the Year*, Hipparchus came to the conclusion that the equinoctial points move at least 1° per century in a direction opposite to the order of the zodiacal signs.[2]

In the one case he implicitly expresses his observation as the 'revolution of the eighth sphere' and on the other hand he refers implicitly to the 'precession of the equinoxes'. For the Indian astronomers it is the latter conception which was familiar to them, i.e. the sphere of the fixed stars was considered fixed and it was the equinoxes that moved. Indian astronomers thus continued the Babylonian tradition. The Greek (Western) tradition, following Hipparchus and Ptolemy, regarded the equinoxes as fixed and endowed the sphere of the fixed stars with a slow forward motion, like a slow-moving planet; it is this conception to which the Platonic world-year tradition discussed by Duhem refers.[3]

The Platonic tradition of a Great Year—as it subsequently became interpreted—required that not only the Grand conjunction of all the planets, but also the eighth sphere (or a specific zodiacal division in the eighth sphere) should return to the vernal point, or one of the tropical points.[4] The Babylonian astronomer Berossus[5] interpreted the Great Year as follows:

> Berossus, who thus interprets the Babylonian tradition, says that those events take place accordingly to the course of the stars; and he affirms it so positively as to fix the time for the Conflagration and the Deluge. He maintains that all terrestrial things will be consumed when the planets, which now are traversing their different courses, shall all coincide in the sign of Cancer, and be so placed that a straight line could pass directly through all their orbs. But the inundation will take place when the same conjunction of the planets shall occur in Capricorn.[6]

If Berossus' statement is interpreted according to the Platonic tradition, then "the period of the Great Year of the Chaldeans conceived of by Berossus [is] the time which separates two successive conjunctions of all the wandering

---

1. Ptolemy, *Almagest* VII, 2 (trsl. Toomer, p327).
2. Neugebauer, HAMA I, p293, quoting Ptolemy from *Almagest* VII, 2.
3. Duhem, *Le système du monde* I, pp65–85, pp284–296; II, pp214–223, gives an account of this tradition *as it subsequently became interpreted*, i.e. not necessarily corresponding to the views of Plato himself. We can only conjecture what Plato's view was (if he had one at all).
4. Joannis Stobaei, *Eclogarum physicarum* I, 8 (cf. Duhem, *Le système du monde* I, pp72–23) requires the Grand Conjunctions to recur at the place of the summer solstice.
5. Bidez, *Bérose et la grand-année*.
6. Seneca, *Naturalium Quaestionum* III, 29 (trsl. Cory, p328).

stars at the spring equinoctial point."[1] In the formulation of Berossus, interpreted according to the Platonic tradition, Grand Conjunctions occur at the vernal point, but between the first and the second, the eighth sphere, whose movement is measured by the location of the vernal point, rotates through 180°, from Cancer to Capricorn. At the first Grand Conjunction the significator of the eighth sphere, i.e. the location of the vernal point, is Cancer and at the second it is Capricorn, diametrically opposite to Cancer.

Of course, Berossus himself, as a Babylonian, would have had his own formulation of the concept of the Great Year. Our question is: How were schemes such as that of Berossus interpreted in the Indian astronomical tradition? Since there was not this same conception of a rotating eighth sphere in Indian astronomy, Hindu astronomers conceived of this sphere as fixed, treating the regression of the vernal point through the zodiac simply as one of the long-term slowly-changing astronomical variables such as the nodes and apogees of the planets (excluding the lunar nodes, Rahu and Ketu, which were essentially regarded as planets). Thus the *kalpa* system of Indian astronomy reckoned with the changing sidereal locations of the planetary nodes and apogees. Hence the steady retrogression of the vernal point was computed simply as one of the changing astronomical parameters in the later *kalpa* system. The Grand Conjunction determining the world-year in the *kalpa* and *mahayuga* systems had to occur at 0° Aries (sidereal).

Since the vernal point, defined to be 0° Aries in the tropical zodiac, was—at any point in time except for the first few centuries AD—generally far removed from 0° Aries in the sidereal zodiac, a Grand Conjunction of the planets at the vernal point was not only a different conception from that of the Hindu one of the *kalpa* and *mahayuga* systems but also it was different in terms of astronomical practice. Apart from this difference there still remains the question whether the conception of a Great Year—as a basic conception belonging to Indian astronomy—was an adaptation of the "Great Year of the Chaldeans conceived of by Berossus."[2]

HOW THE INDIAN SIDEREAL ZODIAC DIFFERS FROM THAT OF THE BABYLONIANS

The transmission of certain elements of Greco-Babylonian astrology to India can be traced back to the second century AD.[3] It was at this time that the ecliptic coordinate system of the zodiac became introduced into India, probably via

1. Duhem, *Le système du monde* II, p 216.
2. Ibid.
3. Pingree (1978).

Greek influence stemming from Alexandria, the principle center from which astrology spread in antiquity.[1] The astrological doctrines elaborated in Hellenistic Egypt were primarily Greek adaptations of Babylonian ideas.[2] The astronomical techniques employed by the early Greek astrologers were, similarly, adaptations of Babylonian astronomical methods.[3] For example, Greek astrologers used the sidereal zodiac as their coordinate system, thus continuing the Babylonian tradition. This is evident from an analysis of the horoscopes in the *Anthology* of Vettius Valens (second century AD),[4] who was a contemporary of Ptolemy, and is one of the most important sources for our knowledge of Greek astrology. However, there is evidence of inconsistencies in the specification of the sidereal zodiac by different authors, e.g. Pliny (first century AD) referred to a location of the vernal equinox at 8° Aries,[5] whilst horoscopes in Vettius Valens' literary collection place the vernal point at about 5° Aries in AD 50 and 3½° Aries in AD 160.[6] Evidently neither author had a sure definition of the Babylonian sidereal zodiac, since the definition of the Babylonian zodiac in conformity with cuneiform sources indicates that the vernal point was located at about 2½° Aries in AD 40 and 1° Aries in AD 150 (Appendix I).

As the recipients of Greco-Babylonian astrology, Indian astronomers and astrologers inherited the sidereal zodiac at some time around the second century AD, but subsequently redefined it.[7] It was not until the Greek phase of Hindu astronomy, beginning in the fifth century AD, that the sidereal coordinate system of Indian astronomy received a secure definition. Unlike the original Babylonian definition of the zodiac (a star catalogue in which Normal Stars were assigned longitudes in the twelve 30° signs),[8] the Indian definition was purely mathematical-astronomical, as discussed below. The Indian sidereal zodiac was defined theoretically and thus, as far as can be ascertained now, was not placed in relation to the zodiacal fixed stars by means of a star catalogue. Both the sidereal zodiac of the Babylonians (equal division signs) and the astronomical zodiac of the Greeks (unequal division constellations,

---

1. Cf. Pingree (1963) for details of this transmission.
2. Cf. Sachs (1952, 1) for a summary of this early development of astrology.
3. Neugebauer-van Hoesen, *Greek Horoscopes*, pp 170 ff.
4. Ibid., pp 78–131 and pp 176 ff.
5. Pliny, *Naturalis Historia* XVIII, 59 (trsl. Rackham V, p 329/331).
6. Neugebauer-van Hoesen, *Greek Horoscopes*, p 180.
7. Just as at that time Greek astrologers such as Vettius Valens (second century AD), although utilizing the Babylonian sidereal zodiac, did not have a secure definition thereof, so also it seems that Indian astronomers and astrologers also lacked a secure definition of the sidereal zodiac when it was transmitted to India. This is apparent from the fact discussed below that Indian astronomers redefined the sidereal zodiac in the sixth century AD.
8. Sachs (1952, 2); cf. also Appendix I.

compiled by Ptolemy) were defined by star catalogues, in terms of the fixed stars themselves. However, the definition of the Indian sidereal zodiac, as discussed below, was a function of the *yuga* system. If the theoretical definition of the twelve 30° signs of the Indian zodiac had been exact, a direct correspondence with the Babylonian zodiac (Appendix I) would exist and there would be no deviation (in the course of time) between the two zodiacs.[1] The fact is, however, that the definition of the Indian zodiac implicit within the various *yuga* systems is not perfectly exact. The result is that there is a slowly-changing relationship between the Indian and Babylonian sidereal zodiacs, which is noticeable over a long period of time.

How does the discrepancy between the two zodiacs arise? It follows from the *a priori* assumption of the *yuga* system that a Grand Conjunction of all the planets took place at 0° Aries in −3101. Indian astronomers employing the *yuga* system computed planetary sidereal longitudes by counting the planetary revolutions measured from 0° Aries since −3101. The accuracy of their computations was dependent upon the accuracy of the astronomical parameters used. In particular, the accuracy of the Indian definition of the zodiac depended on the accuracy of the parameter giving the sidereal period of the sun, since the Indian zodiac was defined in terms of the longitude of the sun by "the actual completion of so many degrees by the sun, called *sankranti*, where the number of degrees is computed by *rasis*, i.e. in multiples of 30 degrees."[2]

The first *rasi*, when the sun's sidereal longitude was 0°–30°, was the sign of Aries (Mesha); the second rasi, when the sun's sidereal longitude was 30°–60°, was the sign of Taurus (Rishabha); etc. The determination of the Indian zodiac thus relied upon the sidereal longitude of the sun, i.e. upon the sun's sidereal period as computed from the astronomical parameters inherent in whichever *yuga* system was employed. The zero point of the Indian zodiac was therefore mathematically defined as the moment when a computed sidereal revolution of the sun was completed. In theory 0° Aries (Indian) was fixed in the heavens (fixed in relation to the fixed stars comprising the zodiacal belt), but because of slight differences in each of the various *yuga* systems between the actual sidereal period of the sun and its computed value, the zero point moved slowly in relation to the fixed stars. For example, the sidereal period of the sun as computed according to the (new) *Surya siddhanta*, the most widely used treatise expounding the *yuga* system, differed from the actual sidereal period by 0.002361° per year.

---

1. Except for the minute deviation arising by virtue of the proper motion of the fixed stars.
2. Pillai, *An Indian Ephemeris* 1, p 6.

134 HISTORY OF THE ZODIAC

[The] 0° point of Indian celestial longitude, which in theory is supposed to be absolutely fixed in the heavens, is actually shifted every year, so far as the sun is concerned, by 0.002361 of a degree. For 1400 years this shift amounts to 3.3054 or a little less than 3⅓ degrees (following the length of the sidereal year as given in the *Surya Siddhanta*). If, according to the general opinion of historians of Indian astronomy, the sun was at the zero point of longitude and in Zeta Piscium on or about 18 March AD 532, this present zero point in AD 1932 is, according to Indian astronomy, 3⅓ degrees to the east of Zeta Piscium.[1]

Figure 19

*The Sidereal Motion of the Zero Point,
0° Aries, of the Indian Zodiac*

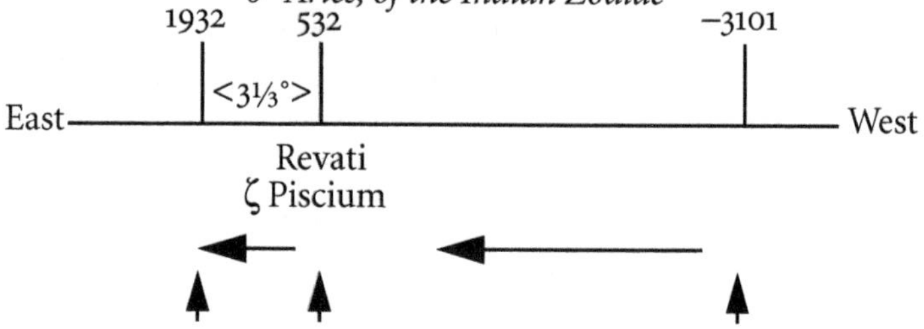

```
           1932   532                         -3101
East ───────┤<3⅓°>├─────────────────────────────┤──── West
            Revati
            ζ Piscium
```

Movement of the zero point of the Indian zodiac in relation to ζ Piscium.

Between AD 532 and 1932 the zero point 0° Aries (Indian) shifted some 3⅓° degrees east of ζ Piscium (known to the Indians as Revati). Correspondingly, in −3101 the zero point of the Indian zodiac lay approximately 8½° degrees west of Revati (Figure 19) since, in the intervening 3633 years, 0° Aries (Indian) shifted (in relation to the fixed stars) about 8½° at a rate of 0.002361° a year. Revati has a slight proper motion itself, but to the nearest degree it has been located at 25° Pisces in the Babylonian sidereal zodiac throughout the period beginning in −3101 (Appendix I). The motion of 0° Aries (Indian) can be computed accurately in relation to the Babylonian zodiac, since the latter is truly fixed in relation to the fixed stars, defined in relation to Aldebaran at 15° Taurus and Antares at 15° Scorpio (Appendix I).[2]

From recent studies it is evident that 0° Aries (Indian) was located 0°10' east

---

1. Ibid., pp 90–91.
2. Disregarding the very slight motion of the Babylonian sidereal zodiac given by the proper motion of the stars Aldebaran and Antares (see Appendix I).

of ζ Piscium in AD 562/3.¹ This location of the zero point by the Indians was adopted by the Islamic astronomer al-Khwārizmi (first half of ninth century) in the Khwārizmian Tables, and by the Jewish astronomers who prepared the Toledan Tables (eleventh century).² AD 562/3 is the date at which the vernal point coincided with the zero point of the Indian zodiac.³ Allowing a rate of precession of 1° in 72 years, the coincidence of the vernal point with 0° Aries (Babylonian) occurred in about AD 220.⁴ In the period of 342 years between AD 220 and AD 562, the vernal point retrogressed 4°45' (72 x 4¾ = 342). Thus, assuming the accuracy of the reconstructed Babylonian zodiac in Appendix I, the difference between 0° Aries (Indian) and 0° Aries (Babylonian) in AD 562 was 4°45'. This is confirmed by the fact that, according to Mercier,⁵ in AD 562 0° Aries (Indian) was located 0°10' east of ζ Piscium, whose longitude in the Babylonian zodiac (for the epoch AD 220) was 25°05' Pisces. (25°05' + 0°10' + 4°45' = 30°00' equating with 0° Aries.

Since AD 562, 0° Aries (Indian) has been moving steadily east of ζ Piscium towards 0° Aries (Babylonian) at a rate of 0.002361° per year, the rate referred to above. Theoretically, then, in about AD 2578 the two zero points would coincide.⁶ Similarly, in –3101 the zero point of the Indian zodiac must have been located 13°24' west of the zero point of the Babylonian zodiac (Figure 20).⁷

Figure 20

*The Motion of 0° Aries (Indian)
in Relation to the Babylonian Zodiac*

1. Mercier (1976); cf. also Billard, *L'astronomie indienne*, p 117.
2. Mercier (1976).
3. Ibid.
4. According to Huber (1958) in –100 the vernal point lay at 4°28' Aries in the Babylonian zodiac, hence 320 or 321 years later, in AD 220/221, the vernal point lay at 0° Aries. 4°28' = 4.46° and 72 x 4.46 = 321—see Appendix I.
5. Mercier (1976).
6. 0.002361° per year corresponds to 4°45' in 2016 years, and 562 + 2016 = AD 2578.
7. 0.002361° per year corresponds to 13°24' in 5679 years, and AD 2578 – 5679 = –3101.

136 HISTORY OF THE ZODIAC

The *yuga* system of Indian astronomy is based on the *a priori* assumption that there was a Grand Conjunction of the planets at 0° Aries (Indian sidereal) in −3101. Assuming that such a conjunction did actually take place on 17/18 February −3101,[1] then the point which in Indian astronomy is defined to be 0° Aries, where the mean conjunction of the planets occurred, is in the Babylonian definition 16°36' Pisces (Figure 20). Thus a discrepancy between the Indian and Babylonian zodiacs was inherent in the *yuga* system from the beginning. However, the inaccuracy of the parameter for the sidereal period of the sun in the (new) Surya siddhanta would eventually compensate for the discrepancy between the Indian and Babylonian zodiacs. In about AD 2578, through the compensatory motion of 0° Aries (Indian), there would be no discrepancy between them and the zero points of the two zodiacs would coincide, if the inaccurate parameter for the sun's sidereal period were to continue to be used.

A NEW DEFINITION
OF THE INDIAN SIDEREAL ZODIAC

Reviewing the history of the zodiac in India: during the fifth/sixth centuries AD there took place a 'revolution' in Indian astronomy through which the sidereal zodiac in India became defined in a unique way. A central figure in the reformation that occurred in Indian astronomy is Aryabhata, who was born in AD 476. His astronomical system took as epoch date the year AD 499, and is based on the computation of large numbers of planetary revolutions of the zodiac over long periods of time (the *yuga* system). The reform in Indian astronomy that occurred at that time through Aryabhata resulted in a definition of the Indian zodiac different from that of the Babylonian zodiac, recalling, as discussed above, that the Babylonian zodiac had probably been introduced into India in the second century AD. This new and original definition of the Indian zodiac was evidently necessitated by the fact that no one knew exactly how the Babylonian zodiac was defined in relation to the fixed stars. Thus a new sidereal zodiac, the Indian zodiac, came about as a division of the zodiacal belt into twelve 30° signs, with the starting point of the sign of Aries taken to be the fixed star that was *at that time* (shortly after the *yuga* system was introduced) coinciding with the location of the vernal point. The

---

1. Cf. Pingree (1976), pp115–116, for a computation of planetary positions at sunrise on 18 February −3101 using Ptolemy's parameters from the *Almagest*. The result shows that (depending upon the reliability of the parameters) there was a loose conjunction of the planets and the lunar descending node in the vicinity of ζ Piscium; cf. also van der Waerden (1970), p46 for a computation using modern tables.

fixed star nearest to the vernal point, at the time the reform was finalized around the middle of the sixth century AD, was ζ Piscium (known in India as *Revati*). Hence the Indian zodiac defined in the sixth century AD is defined so that its zero point (0° Aries) was located 0°10' east of the star Revati (or, according to some authorities, 0° Aries *coincided* with Revati). Thus Colebrooke in his classic study of Indian astronomy wrote:

> All authorities agree that the principle star (Revati) ... has no latitude, and two of them assert no longitude; but some make it ten minutes short of the origin ... (of the Indian zodiac), viz. 359 degrees 50 minutes. This clearly marks the star ... in the string of the Fishes (ζ Piscium); and the ascertainment of it is important in regard to the adjustment of the Hindu sphere.[1]

As referred to above, Mercier's study of the medieval conception of precession leads to the conclusion that 0° Aries of the Indian zodiac was fixed in the years AD 562/3, when the vernal point lay 0°10' east of ζ Piscium.[2] By virtue of its definition the Indian zodiac is a *fixed* zodiac, i.e. it is sidereal, defined in relation to stars (or rather a star) in the zodiacal belt. In this respect it is similar to the Babylonian sidereal zodiac, which is also a fixed star (sidereal) zodiac. In contrast the Greek tropical zodiac, which is defined with the vernal point specified as 0° Aries, is a *moving* zodiac since, on account of precession, the vernal point moves in a retrograde direction in relation to the zodiacal belt. The Greek tropical zodiac coincided with the Babylonian sidereal zodiac about AD 220.[3] Thus, when the zodiac was introduced into India, probably in the second century AD, the Babylonian sidereal zodiac and the Greek tropical zodiac were more or less identical. However, by the time of Aryabhata, a discrepancy between the two existed because of precession. The very nature of Indian astronomy, based on planetary revolutions of the zodiacal belt, demanded a fixed (sidereal) zodiac. With no other definition of a fixed star zodiac at hand (such as that given for the Babylonian zodiac in the Babylonian star catalog reconstructed in Appendix I), a sidereal zodiac had to be newly defined. This was accomplished simply by *fixing* the moving tropical zodiac at a specific point in time, in AD 562/3, in relation to the zodiacal belt. The choice of this particular moment for the definition of the Indian sidereal zodiac meant that the star Revati, the star nearest to the vernal point at that time, became the prime fiducial of the Indian zodiac.

Hence, there came into existence two fixed star (sidereal) zodiacs consisting of twelve equal 30° signs: the Babylonian and the Indian sidereal zodiacs. The

---

1. Colebrooke, *Miscellaneous Essays* II, pp343–344.
2. Mercier (1976).
3. See Appendix I.

Greek tropical zodiac coincided with the Babylonian sidereal zodiac about AD 220, when the vernal point was located at the zero point (0° Aries) of the Babylonian zodiac (Appendix I). Similarly, by virtue of its definition, the Indian zodiac was identical with the Greek tropical zodiac about AD 562/3. As described above, in the 342 years elapsing between 220 and 562, the vernal point retrogressed approximately 4°45', allowing a rate of precession of 1 degree in 72 years.[1] Thus, in its original definition, the Indian zodiac was a fixed (sidereal) zodiac, like the Babylonian zodiac, but with the boundaries of the signs of the Indian sidereal zodiac each lying 4°45' west of the corresponding boundaries of the signs of the Babylonian zodiac.

Although the Indian zodiac was originally defined so that its zero point (0° Aries) more or less coincided with the star Revati, this definition generally has not been adhered to up to the present time, owing to the inaccuracy inherent in this system, as discussed above. Thus the authors of the *panchangas*—modern Indian popular calendars—almost invariably use Spica, and not Revati, as the prime fiducial star. They define the signs of the Indian sidereal zodiac in relation to Spica, so that Spica falls at the boundary between the sidereal signs Virgo and Libra. Thus Spica is defined to be at 30° Virgo (0° Libra). This modern Indian fixed star (sidereal) zodiac, with Spica as the prime fiducial, was officially recognized in 1955 and adopted for calendar purposes by the Council of Scientific and Industrial Research on behalf of the government of India.[2]

Taking Spica to be at 30° Virgo, the modern Indian zodiac is considerably closer to the Babylonian zodiac than its forerunner—the original Indian zodiac defined in the sixth century AD with Revati as the prime fiducial star. Thus the longitude of Spica in the Babylonian sidereal zodiac was 29°07' Virgo for the epoch –100 (Appendix I), signifying a difference of a little less than 1° between the modern Indian zodiac and its Babylonian prototype, as compared with a difference of almost 5° between the Indian zodiac defined in the sixth century AD and the Babylonian zodiac.

In the definition of the modern Indian zodiac, with the adoption of a prominent first magnitude star (Spica) to delineate the end of the sign of Virgo (or

---

1. 4¾ × 72 = 342 years.
2. *Report of the Calendar Reform Committee* (1955). This report made by the Council of Scientific and Industrial Research was commissioned by the government of India, which shows the importance of the zodiac—still to the present day—throughout all levels of society in India. The reason for specifying an "official" zodiac is that many public festivals in the Hindu calendar are determined by astronomical considerations, e.g. the beginning of the religious year is specified by the Sun's entrance into the sign of Aries (in the Indian sidereal zodiac). Thus a precise definition as to when the Sun enters Aries is required, and this depends upon a secure definition of the zodiac and the sign of Aries. This is just one example of many.

the beginning of the sign of Libra), a definite similarity can be seen to exist with the way in which the Babylonian zodiac came into existence. The only difference is that, whereas Babylonian astronomers chose Aldebaran at the center of the sign of Taurus and Antares at the center of the sign of Scorpio, later Indian astronomy chose Spica to be the primary reference star for their definition of the zodiac, with Spica marking the midpoint of the zodiac at the boundary between the signs of Virgo and Libra. What is remarkable, however, is the small difference—only about one degree—in the two zodiacs resulting from these different definitions. This is a consequence of the actual spatial relationship between the stars Aldebaran and Antares on the one hand, and Spica on the other hand, which gives rise to (from the above definitions) two fixed star sidereal zodiacs differing by approximately one degree in the specification of the longitude of their zero point (0° Aries sidereal) and thus yielding a 1° difference in all longitudes.

APPENDIX I

BABYLONIAN STAR CATALOGUE RECONSTRUCTED

THE INTRINSIC DEFINITION OF THE BABYLONIAN ZODIAC

IN THIS THESIS on the definition of the zodiac, as discussed in Chapter 5, it is postulated that the *intrinsic definition* of the Babylonian sidereal zodiac is such that the stars Aldebaran and Antares specify an axis running through the middle of the zodiac, with Aldebaran at 15° Taurus and Antares at 15° Scorpio. All other stellar longitudes were then determined by measuring the distances of stars from Aldebaran or Antares. If this conjecture is correct, it was probably in this way that the world's first star catalogue was compiled.[1]

Unfortunately there are only three stellar longitudes (of a total of seven) retrieved from this Babylonian star catalogue that are indicated securely in terms of their assigned longitudes:

The Root of the Barley-Stalk (?): 16 (Virgo)....
The Southern Balance-Pan: 20 Libra....
The Northern Balance-Pan: 25 Libra. [2]

These three stars are identified as Porrima (γ Virginis), Zubenelgenubi (α Librae) and Zubeneschamali (β Librae).[3] Thus the Babylonians located γ Virginis at 16° Virgo, α Librae at 20° Libra and β Librae at 25° Libra in their star catalogue. In the reconstructed Babylonian sidereal zodiac tabulated below, specified by locating Aldebaran at 15° Taurus and Antares at 15° Scorpio, the longitudes of the two brightest stars in Libra for the epoch –100 (101 BC) are:

α² Librae 20°20' Libra; β Librae 24°38' Libra; and the longitude of γ Virginis is 15°39' Virgo.

---

1. Sachs (1952, 2).
2. Ibid., p146.
3. Hunger-Pingree (1999), p150. Zubenelgenubi is also known as Kiffa Australis and Zubeneschamali is otherwise known as Kiffa Borealis.

Rounded to the nearest degree, these are the values given in the original Babylonian star catalogue, providing further confirmation of the validity of the reconstructed star catalogue tabulated below.

The epoch –100 is chosen in order that the reconstructed Babylonian sidereal zodiac tabulated in this Appendix can be utilized by other researchers when making comparisons with ancient observations. This date, twenty-seven years after the end of the Seleucid Era, denotes the last phase of Babylonian astronomy. It is also this epoch of –100 that was chosen by Huber in his survey referred to below.

As mentioned in Chapter 5, confirmation of the validity of the reconstructed Babylonian sidereal zodiac is given by Huber's survey of Babylonian observations and computations[1] leading to the conclusion that for the epoch –100 the vernal point was located at 4°28' Aries, which fits exactly with the reconstructed Babylonian sidereal zodiac, as discussed below.

Since Aldebaran was located exactly 45° east of the vernal point in AD 220 (corresponding at that unique point in time to a tropical longitude of 15° Taurus), then if the sidereal zodiac is defined with Aldebaran at 15° Taurus, the vernal point, which by definition is at 0° Aries in the tropical zodiac, was obviously located at precisely 0° Aries in the sidereal zodiac in AD 220. This observation is central for all the chronological indications referred to in this thesis. Allowing a rate of precession of 1° in 72 years, going back from AD 220 the vernal point was located at 1° Aries in AD 148, 1°40' Aries in AD 100,[2] 2° Aries in AD 76, 3° Aries in AD 4, 4° Aries in –68, 4°27' Aries in –100,[3] 5° Aries in –140, 6° Aries in –212, 7° Aries in –284, 8° Aries in –356, 9° Aries in –428, and 10° Aries in –500.[4] These dates are listed here for the sake of convenience, since they are often referred to in the various chapters of this thesis and also in Appendix II.

BABYLONIAN STAR CATALOGUE

The natural reference system for the motion of the planets, moon and sun, are the fixed stars....[5]

The Babylonian star catalogue of this appendix, tabulated below, is—in terms of its structure—identical to Ptolemy's catalogue of stars, except that

---

1. Huber (1958).
2. The epoch date of *Ptolemy's Catalogue of Stars* compiled by Peters and Knoble.
3. Agreeing with Huber (1958), who determined that the vernal point in the Babylonian sidereal zodiac was located at 4°28' Aries in –100.
4. Agreeing with Kugler, SSB I, p205, with van der Waerden, SA II, p266, and also with Britton-Walker (1996), p49.
5. Neugebauer, HAMA II, p1026.

the longitudes of the stars are given in relation to the Babylonian sidereal zodiac.[1] Peters and Knobel published *Ptolemy's Catalogue of Stars* in 1915.[2] In order to identify the stars catalogued by Ptolemy, Peters and Knobel computed the latitudes and longitudes of all of Ptolemy's stars using for this purpose Piazzi's catalogue[3] reduced to the epoch of AD 100, which is close to the epoch of Ptolemy's catalogue, AD 138.[4] Thus all stellar positions in *Ptolemy's Catalogue of Stars* were derived by Peters and Knobel from Piazzi's catalogue, with the exception of a few stars in Danckwortt's catalogue[5] and some in the catalogue of P.V. Neugebauer,[6] which they reduced (to the epoch of AD 100) from those catalogues respectively.

> The Catalogue of Stars contained in the seventh and eighth books of Ptolemy's *Syntaxis Mathematica*, commonly called the *Almagest*, must always be considered of unique interest. It is the first and most ancient document we possess which gives a description of the heavens of sufficient exactness to admit of comparison with modern observations.[7]

More recently, in his translation of Ptolemy's *Almagest*, G.J. Toomer presents Ptolemy's star catalogue anew in English translation. As Toomer says: "The reader who wishes to go further must still consult Peters-Knobel, on which I have drawn heavily, and which is still the best treatment of the catalogue as a whole, though badly in need of revision and updating in certain respects."[8] The stars listed in the Babylonian star catalogue tabulated below are tabulated in the same way as Catalogue II in *Ptolemy's Catalogue of Stars*,[9] by Peters and Knobel, generally accepting their identification of the 1022 star from Ptolemy's catalogue.[10] This reconstructed Babylonian star catalogue follows

---

1. Ptolemy, *Almagest* VII, 5 and VIII, 1 (trsl. Toomer, pp 341–399).
2. Peters-Knobel, *Ptolemy's Catalogue of Stars*.
3. Piazzi's catalogue.
4. Neugebauer, HAMA I, p 284 points out that although the epoch for Ptolemy's catalogue of stars was AD 138, owing to an error in Ptolemy's specification of the location of the vernal point, the stellar longitudes given in this catalogue in terms of the tropical zodiac (measured from the vernal point) fit better to the epoch AD 48.
5. Danckwortt's catalogue.
6. P.V. Neugebauer's catalogue.
7. Peters-Knobel, *Ptolemy's Catalogue of Stars*, p 7.
8. Toomer (1984), p 14.
9. Ibid., pp 51–73.
10. There are some problems with identifying the stars on account of Ptolemy's verbal description, which is not always clear. Thus Peters and Knobel's reconstruction contains uncertain identifications of stars in 114 cases; cf. Toomer (1984), pp 14–17; Graßhoff (1990), pp 217–269; Strano (1999); and Pickering (2002). Toomer generally follows Peters and Knobel, as does the tabulation below. In the event of a discrepancy, if Toomer is followed instead of Peters and Knobel, this is referred to in the notes beneath the tabulation of individual constellations.

APPENDIX I 143

the model of *Ptolemy's Catalogue of Stars* in its tabulation. However, the stellar longitudes, latitudes, and magnitudes listed in the Babylonian star catalogue were computed or simply taken from the Hipparcos catalogue, reduced to the epoch –100.[1] Assuming that at this epoch Aldebaran's longitude was exactly 15°00' Taurus, it transpires that the longitude of Antares was 15°01' Scorpio for the epoch –100 (Table 7, left hand column). Thus, in the Babylonian star catalogue tabulated below, the longitudes, latitudes, and magnitudes given in *Ptolemy's Catalogue of Stars* compiled by Peters and Knobel have been updated using the modern data provided by the Hipparcos catalogue to the specification of the Babylonian sidereal zodiac with Aldebaran at 15° Taurus and Antares at 15° Scorpio.

Whereas almost all star catalogues must be recomputed every year or every few years because of the shifting position of the vernal point, this reconstruction of Ptolemy's catalogue in terms of Babylonian sidereal longitudes and ecliptic latitudes is valid for thousands of years, apart from the small deviations arising through the proper motion of fixed stars. A similar catalogue of stars, based on the same principle as the Babylonian star catalogue, is that of Copernicus, who catalogued the stars in terms of the fixed stars themselves, i.e. independently of the vernal point. Copernicus also adapted his catalogue from that of Ptolemy. The zero point chosen by Copernicus was the star Mesartim (γ Arietis). Thus he reduced all of the star longitudes in Ptolemy's catalogue by 6⅔° (since Ptolemy placed Mesartim at 6⅔° Aries) to arrive at his catalogue.[2] The reconstruction tabulated below of Ptolemy's catalogue in terms of Babylonian sidereal longitudes is thus essentially an updated version of Copernicus' star catalogue.[3]

---

1. The Hipparcos satellite was launched by the European Space Agency in 1989. It transmitted data until 1993, gathering remarkably accurate information on nearly 118,000 stars, measuring their positions, magnitudes, and parallaxes with far greater accuracy than any previous study using telescopes on the earth. Toomer's remark is relevant here: "I have made no attempt to redo the work of Peters and Knobel, namely to compute the longitude and latitude of the relevant stars for Ptolemy's time from modern data (in particular using the most up-to-date values for the proper motions). This might be worth while, though I doubt whether the degree of improvement would justify the large amount of computation."(1984), p16. I can confirm this remark by G.J. Toomer. Comparison of the values (tabulated below) obtained via the Hipparcos catalogue, providing the most up-to-date values, generally differ only minimally from the values obtained via Peters and Knobel. The difference between the Hipparcos catalogue and that of Peters and Knobel in the values obtained for longitudes and latitudes—very often there is no difference at all—usually does not exceed one minute of arc for most stars. Cf. also p146,n1 below.

2. Copernicus, *De revolutionibus orbium coelestium* II, 14 (trsl. Rosen, pp 85–117).

3. Since Mesartim's longitude in Copernicus' star catalogue is 0° Aries and since it is 8½° Aries in the reconstructed Babylonian star catalogue (see below), the two catalogues—identical

In this way the most outstanding of astronomers [Ptolemy] noted the distance of each of the stars from the vernal equinox at that time, and he set forth the constellations of the celestial creatures. By these achievements he gave no small assistance to this study of mine and relieved me of quite an arduous task. I believed that the places of the stars should not be located with reference to the equinoxes, which shift in the course of time, but that the equinoxes should be located with reference to the sphere of the fixed stars. Hence I can easily start the cataloguing of the stars at some other unchangeable beginning. I have decided to commence with the Ram, as the first zodiacal sign, and with its first star [Mesartim], which is in its head. My purpose is that in this way always the same definitive appearance will remain for those bodies which shine as a team, as though fixed and linked together, once they have taken their permanent place.[1]

The standpoint adopted by Copernicus is exactly that which the Babylonians had taken some 17 or 18 centuries before him. In the Babylonian lunar systems (A and B in Neugebauer's terminology[2]) the equinoxes were referred to the fixed-star zodiac. In system A the vernal point was placed at 10° Aries, and in system B at 8° Aries.[3] Copernicus thought that the vernal equinox should be referred to the sphere of the fixed stars. The Babylonians actually did refer the vernal point to this sphere, which they divided into twelve equal signs. Indeed, from the point of view of simplicity the Babylonian system has the advantage that all the stellar longitudes and latitudes are given with the effect of precession removed, and the only variable that needs regular recalculation is the location of the vernal point. Over long periods of time the proper motions of stars needs to be taken into account, leading to small changes in the longitudes and latitudes of stars, usually amounting to substantially less than ¼° in longitude (and, similarly, less than ¼° in latitude) in the space of one thousand years, although there are a few exceptions to this, particularly in the case of stars that are relatively close to our solar system. In contrast, if stellar locations are referred to the vernal point, as is customary in modern astronomy (using

---

in terms of structure, as both are adapted from Ptolemy's catalogue—are completely equivalent to one another, with a constant difference in stellar longitudes amounting to 8½°. To transform Copernicus' catalogue to the Babylonian norm, 8½° would have to be added to all longitudes in his star catalogue. Another example of a star catalogue following the same principle as that of the Babylonian star catalogue is provided by Ptolemy himself in his *Handy Tables*, where he tabulated the positions of about 180 bright stars within the zodiacal belt, i.e. with latitudes between 10° north and 10° south. The tabulation of zodiacal stars in the *Handy Tables* is based on the catalogue of the *Almagest* but with the star Regulus serving as the zero point from which the longitudes of the other zodiacal stars are measured. These are therefore ecliptic sidereal longitudes measured from Regulus at 0°.

1. Copernicus, *De revolutionibus orbium coelestium* II, 14 (trsl. Rosen, pp 83–84).
2. Neugebauer, ACT I, p 41.
3. Ibid., p 72.

right ascension and declination), then all the stellar locations need to be recalculated regularly every few years in relation to the vernal point.

The difference between Copernicus' sidereal catalogue and that of the Babylonians lies in the choice of the primary reference (fiducial) star(s). Copernicus selected Mesartim as the zero point for his catalogue and, as referred to in footnote 2 on the previous page, Ptolemy chose Regulus as the zero point for his catalogue in the *Handy Tables*. It is not known how the Babylonians arrived at zodiacal longitudes for Normal Stars, but in conformity with the reconstructed Babylonian star catalogue the two diametrically opposed first magnitude stars Aldebaran (α Tauri) and Antares (α Scorpii) are located at 15° Taurus and 15° Scorpio, respectively. Later Cleomedes, the author of an astronomical textbook and possibly the recipient (via Posidonius) of Babylonian astronomy and astrology, also assigned these two stars stellar longitudes in the middle of their respective signs.[1] Thus it may be conjectured that Aldebaran at 15° Taurus and Antares at 15° Scorpio provided the fiducial axis for the Babylonian zodiac.

Once a fiducial star (or axis) has been chosen, all stellar longitudes may be referred to it. Apart from proper motions "the same definitive appearance will remain for those bodies [of the stars]."[2] Moreover, the displacement due to proper motion is very slight.

> For historical purposes proper motion will rarely be of significance. There are only very few bright stars which show displacements of more than 1 minute of arc per century; Figure 15 [21] illustrates the change of position of Procyon and of Sirius from the time of Hipparchus (black dots) to modern times (white dots) in relation to Orion (for which proper motions cannot be shown in the scale of our diagram).[3]

It can be seen that the distance moved by these fixed stars, even over two millennia, is very small (Figure 21). Between −100 and the present time (epoch 2000), i.e. in the space of 2100 years, Sirius' and Procyon's sidereal longitudes have decreased by 33' and 19', and their latitudes have changed by 57' and 24', respectively. These values expressing the proper motion of the two stars are relatively high, as Sirius and Procyon are very close to our solar system, being located only 9 and 11 light years away, respectively. Most stars are considerably further away than our close neighbors Sirius and Procyon, and so the average difference owing to proper motion between −100 and the present time for most other stars, derived from fifty examples considered below, is about 0°12'

---

1. Neugebauer, HAMA I, p960. See discussion of Cleomedes in Chapter 5.
2. Copernicus, *De revolutionibus orbium coelestium* II, 14 (trsl. Rosen, p84).
3. Neugebauer, HAMA III, p1085, where Neugebauer's figure 15 is reproduced here as figure 21 (see page 147).

of arc in longitude and about 0°12' in latitude. In fact, if sidereal longitudes are expressed in terms of the Babylonian zodiac with Aldebaran fixed at 15° Taurus, while the latitudinal component of proper motion, averaging 0°12', remains the same, the average longitudinal component of proper motion is reduced to less than 0°04', because Aldebaran's proper motion in longitude since –100 amounts to about 0°08' and this value is subtracted from the shift due to proper motion for all other stars.

To gain a clearer idea based on examples of the proper motion of a number of stars, let us consider Table 7 in which the latitudes and sidereal longitudes ('Babylonian style' with Aldebaran fixed at 15° Taurus) of the fifty brightest stars within ± 8° of the ecliptic are given. In the left-hand column are listed the (Babylonian) catalogue values (epoch –100), and in the right-hand column are given the modern (Babylonian) values referred to the epoch 2000.[1]

Neugebauer's Figure 15 is reproduced here as Figure 21, where the black dots indicate the positions for Sirius and Procyon in Hipparchus' day and the white dots indicate the present positions.

---

1. All the stellar longitudes and latitudes listed in Table 7 are sidereal ('Babylonian style') with Aldebaran fixed at 15° Taurus. They are derived from the Hipparcos star catalogue for the epoch AD 2000 (see right-hand column), whereby the figures in the left-hand column have been reduced (allowing for proper motion) to the epoch –100. As with the tabulation of the reconstructed Babylonian sidereal zodiac, Table 7 was computed using Peter Treadgold's Astrofire program that utilizes stellar data from the Hipparcos catalogue.

As an independent check on the accuracy of the data, a comparison of the stellar longitudinal values in the left-hand column of Table 7 was made with the figures published in 1915 by Peters and Knobel. This check reveals the following. Of the 50 stars listed in Table 7, 49 are also listed in the reconstructed Babylonian star catalogue below (the star Electra belonging to the Pleiades is missing). Of these 49 stars, 24 of the stellar longitudes are identical to those computed by Peters and Knobel; in 22 cases there is a difference in stellar longitude of one minute of arc; in 1 case the difference amounts to two minutes of arc; in 1 case it is three minutes of arc; and in 1 case the difference is six minutes of arc probably on account of a computational error. It has to be borne in mind that for this comparison the tropical ecliptic longitudes published in 1915 by Peters and Knobel were converted to sidereal ecliptic longitudes ('Babylonian style') with Aldebaran fixed at 15° Taurus by simply adding 2⅓° to the stellar longitudes given by Peters and Knobel, since they give Aldebaran's tropical longitude (epoch AD 100) to be 12⅔° Taurus. Further, as the epoch chosen by Peters and Knobel is AD 100, some of the differences in sidereal longitudes of stars (from these 49 cases) could be due to the proper motion of stars between -100 and +100. On the whole most of the stellar longitudes differ hardly at all from the values derived from the Hipparcos catalogue.

Figure 21

Proper Motions of Procyon and Sirius since the time of Hipparchus

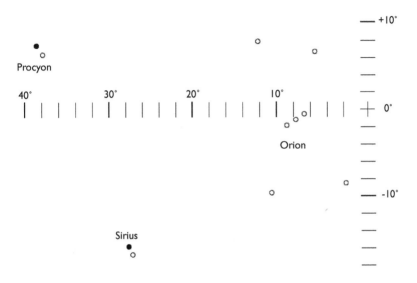

Both columns of Table 7 refer to the fiducial axis Aldebaran-Antares (15° Taurus–15° Scorpio), although owing to a slight displacement of 0°03' arising through their relative proper motions, if Aldebaran is taken now to be at 15°00' Taurus exactly, then Antares is presently at 14°58' Scorpio as comp^{Procyon} th 15°01' Scorpio in –100 (Table 7). In terms of longitudinal motion the fastest-moving star in Table 7 is Zavijava (β Virginis), which in 2100 years has moved 0°26' in longitude (but only 0°03' in latitude). More typically, since –100 Spica (α Virginis) has moved 0°04' in longitude and 0°06' in latitude and Antares (α Scorpii) has moved 0°03' in longitude and 0°15' in latitude. In fact, for the stars listed in Table 7 the average rate of proper motion in longitude during this time interval of 2100 years is less than 0°04' and in latitude is about 0°12'.[1] Thus the sidereal longitudes and latitudes of the stars (measured 'Babylonian style') are essentially the same now as they were in –100.

---

1. Here the proper motion in latitude is larger than in longitude since the stellar longitudes are all measured relative to Aldebaran at 15° Taurus. Aldebaran's proper motion in longitude during this period of 2100 years amounted to 0°08', accounting for the difference.

Table 7
The Fifty Brightest Zodiacal Stars (within lat. ± 8°)

CAT.		LONG. [−100]	LAT. [−100]	NAME	LONG. [+2000]	LAT. [+2000]
η	Psc	2 ♈ 04	+5; 16	Alpherg	2 ♈ 02	+5; 23
17	Tau	4 ♉ 39	+3; 59	Electra	4 ♉ 37	+4; 11
η	Tau	5 ♉ 14	+3; 50	Alcyone	5 ♉ 12	+4; 03
27	Tau	5 ♉ 36	+3; 42	Atlas	5 ♉ 34	+3; 55
θ²	Tau	13 ♉ 08	−6; 04		13 ♉ 10	−5; 50
ε	Tau	13 ♉ 39	−2; 48	Ain	13 ♉ 41	−2; 34
α	Tau	15 ♉ 00	−5; 37	Aldebaran	15 ♉ 00	−5; 28
β	Tau	27 ♉ 49	+5; 13	Elnath	27 ♉ 47	+5; 23
ζ	Tau	0 ♊ 01	−2; 28	Alhecka	0 ♊ 00	−2; 12
η	Gem	8 ♊ 42	−1; 09	Propus	8 ♊ 39	−0; 53
μ	Gem	10 ♊ 31	−1; 02	Tejat Posterior	10 ♊ 31	−0; 49
γ	Gem	14 ♊ 19	−7; 00	Alhena	14 ♊ 19	−6; 45
ε	Gem	15 ♊ 11	+1; 48	Mebsuta	15 ♊ 09	+2; 04
δ	Gem	23 ♊ 46	−0; 26	Wasat	23 ♊ 44	−0; 11
λ	Gem	24 ♊ 02	−5; 52		23 ♊ 59	−5; 38
β	Gem	28 ♊ 45	+6; 31	Pollux	28 ♊ 26	+6; 41
κ	Gem	28 ♊ 55	+2; 51		28 ♊ 53	+3; 05
o	Leo	29 ♋ 34	−3; 52	Subra	29 ♋ 27	−3; 45
η	Leo	3 ♌ 08	+4; 43		3 ♌ 07	+4; 52
α	Leo	5 ♌ 12	+0; 22	Regulus	5 ♌ 02	+0; 28
ρ	Leo	11 ♌ 38	+0; 02		11 ♌ 36	+0; 09
β	Vir	1 ♍ 56	+0; 39	Zavijava	2 ♍ 22	+0; 42
γ	Vir	15 ♍ 39	+2; 56	Porrima	15 ♍ 21	+2; 47
α	Vir	29 ♍ 07	−1; 57	Spica	29 ♍ 03	−2; 03
α²	Lib	20 ♎ 20	+0; 32	Zubenelgenubi	20 ♎ 18	+0; 20
σ	Lib	26 ♎ 00	−7; 27	Brachium	25 ♎ 54	−7; 39
δ	Sco	7 ♏ 49	−1; 45	Dschubba	7 ♏ 47	−1; 59
π	Sco	8 ♏ 12	−5; 14		8 ♏ 09	−5; 29
β	Sco	8 ♏ 26	+1; 15	Graffias	8 ♏ 24	+1; 00
σ	Sco	13 ♏ 03	−3; 47	Al Niyat	13 ♏ 01	−4; 02
α	Sco	15 ♏ 01	−4; 19	Antares	14 ♏ 58	−4; 34
τ	Sco	16 ♏ 43	−5; 52		16 ♏ 40	−6; 07
η	Oph	23 ♏ 11	+7; 32	Sabik	23 ♏ 11	+7; 12
θ	Oph	26 ♏ 38	−1; 35		26 ♏ 36	−1; 51
ξ	Ser	29 ♏ 48	+8; 11		29 ♏ 45	+7; 56
γ²	Sgr	6 ♐ 32	−6; 49	Alnasl	6 ♐ 28	−6; 59
δ	Sgr	9 ♐ 48	−6; 13	Kaus Media	9 ♐ 47	−6; 28
λ	Sgr	11 ♐ 34	−1; 58	Kaus Borealis	11 ♐ 32	−2; 08
ø	Sgr	15 ♐ 23	−3; 41		15 ♐ 23	−3; 57
σ	Sgr	17 ♐ 37	−3; 13	Nunki	17 ♐ 36	−3; 27
ξ²	Sgr	18 ♐ 40	+1; 56		18 ♐ 40	+1; 40

CAT.		LONG.	LAT.	NAME	LONG.	LAT.
ζ	Sgr	18♐53	−6; 56	Ascella	18♐51	−7; 11
τ	Sgr	20♐05	−4; 58		20♐03	−5; 05
π	Sgr	21♐29	+1; 41	Albaldah	21♐28	+1; 26
α²	Cap	9♑05	+7; 10	Algedi	9♑04	+6; 56
β	Cap	9♑17	+4; 50	Dabih	9♑15	+4; 35
ε	Aqr	16♑58	+8; 16	Albali	16♑56	+8; 05
γ	Cap	26♑55	−2; 22	Nashira	27♑00	−2; 33
δ	Cap	28♑35	−2; 33	Deneb Algedi	28♑45	−2; 36
λ	Aqr	16♒49	−0; 17		16♒47	−0; 23

In the reconstructed Babylonian star catalogue tabulated below, the Latin name of each of the 48 constellations from Ptolemy's *Almagest* is given, together with an appropriate English designation of the original Greek description of the figure formed by the stars in each constellation.[1] Altogether there are 21 northern constellations, 12 zodiacal constellations, and 15 southern constellations. Ptolemy's verbal description of the location of each star is included in the tabulation,[2] and also the star name, if known.[3]

The first column gives the modern catalogue designation of each star;[4] the second, the longitude in degrees and minutes in the relevant sign of the Babylonian sidereal zodiac computed for the epoch −100;[5] the third, the latitude (+ north, − south) in degrees and minutes north or south of the ecliptic

---

1. Cf. Allen, *Star Names*. For example, Ptolemy refers to the "Constellation of the Horse". In his description of the figure formed by the stars, the reference to wings makes it clear that he is referring to Pegasus, and so this constellation is identified in the tabulation below as PEGASUS (Constellation of the Winged Horse).

2. Ptolemy, *Almagest* VII, 5 and VIII, 1 (trsl. Taliaferro, pp 234–258), modified, where necessary for the sake of accuracy, by Toomer's translation (1984), pp 341–399. Cf. also p 150, n 2.

3. Allen, *Star Names*; cf. also Kunitzsch, *Arabische Sternnamen in Europa*, Kunitzsch, *Der Almagest*, and Barton and Barton, *A Guide to the Constellations*, pp 72–73.

4. Greek letter classification of the stars, usually in descending order of apparent brightness, was introduced by Bayer in the *Uranometria* (1603) and the numbering system, based on the order of right ascension, was introduced by Flamsteed in the early 1700s. When the Bayer letters ran out (that is, the letters of the Greek alphabet had been exhausted), remaining stars in a constellation were given Roman letters, first a, b, c, and so on, then A, B, C, and so on. In the Babylonian star catalogue each star is usually identified by its Flamsteed number and a Greek or Roman letter. Certain fainter stars, if they have no Flamsteed number or Greek or Roman letter, are identified by Peters and Knobel by their designation (based on subsets of Roman numerals) in the Piazzi catalogue; and otherwise they are identified by their number in the Yale Bright Star Catalogue (BSC), generally following Toomer (1984).

5. The entries in the second column are sidereal longitudes ("Babylonian style" with Aldebaran fixed at ♉15°) computed from the Hipparcos catalogue reduced to the epoch −100, i.e. taking account of the proper motions of all the stars, using Peter Treadgold's Astrofire program.

computed for the epoch −100; the fourth column lists the magnitude;[1] the fifth, the name of the star (if known); and the sixth gives Ptolemy's description of the star's placing.[2]

As mentioned in footnote 10 on page 142, the identification of some of the stars in Ptolemy's catalogue—probably in about 20 cases—is not always certain. Further, some of the stars that Ptolemy places in a given constellation have since been located in a different constellation (identified in the tabulation below),[3] and there are a number of stars listed by Ptolemy as being *unfigured stars about* (or *above* or *below*) a particular constellation (also identified in the tabulation below).[4] Moreover, some of the stars referred to by Ptolemy have since been identified as double stars; if appropriate, both are listed in the following tabulation.[5] Further, Ptolemy's descriptions of stellar positions is not necessarily in the longitudinal order of the stars, since Ptolemy was concerned with describing the stars in each constellation one by one in terms of the *figures* (images) formed by the stars—the *unfigured* stars being those that do not fit into the *figure* (image) of a given constellation. Thus in Ptolemy's

---

1. The magnitudes are taken from the Hipparcos catalogue, in which the stellar magnitudes were accurately measured by the Hipparcos satellite during the years 1989–1993.

2. Toomer (1984), pp341–399, gives the most accurate rendition in English of Ptolemy's description of each star's placing. However, Ptolemy was not very 'reader friendly' in his description. *Ptolemy: The Almagest* translated by R.C. Taliaferro (1952) offers a more 'reader friendly' English rendition, which has been utilized in the tabulation below (modified, where necessary for the sake of accuracy, by Toomer's English rendition of the descriptions). The use of < > indicates Toomer's rendition. There is a consistent difference between Taliaferro's and Toomer's translation to be noted, which concerns the use of the word 'in' or 'on'. For example, the star Mesartim (γ Arieti), one of the stars marking the western horn of the Ram, is "in the horn" according to Taliaferro and "on the horn" according to Toomer. The main difference concerns the use of Ptolemy's 'in advance of' for 'west' and 'to the rear' for 'east'. Thus, for κ Geminorum Toomer translates: "The star on the rear shoulder of the rear twin", whereas Taliaferro translates this as: "The star in the eastern shoulder of the eastern Twin." Similarly, for κ Aurigae (listed by Ptolemy under the unfigured stars around Gemini) Toomer translates: "The bright star in advance of the advance knee", whereas Taliaferro translates this as: "The bright western star of the western knee."

3. For example, the stars now recognized as belonging to the constellation of Columba the Dove were placed by Ptolemy in the south-western outlying part of Canis Major. Further, there are two stars (1*o* and 2η) listed under Boötes which are now regarded as belonging to the Northern Crown (Corona Borealis). These are the stars *o* Coronae Borealis and η Coronae Borealis, which Flamsteed lists as stars number 1 and 2 in the constellation of the Northern Crown.

4. For example, the two stars listed by Peters and Knobel under Perseus with the numbers 14H and 21H are not now regarded as belonging to Perseus. H = Hevelius and these designations refer to the 14th and 21st stars listed under Camelopardalis in Flamsteed's edition of Hevelius' catalogue.

5. Pairs of stars are often designated with *prior* and *posterior* (*post.*), and also with *borealis* ("northern") and *australis* ("southern"), abbreviated as *bor.* and *aus.* Other abbreviations used: *prim.* for *primus* ("first"), *sec.* for *secundus* ("second") and *tert.* for *tertius* ("third").

catalogue the position of each star in the figure of the constellation takes priority over its longitudinal position as measured from the zero point (♈0°) of the zodiac.

BABYLONIAN STAR CALENDAR
(Epoch −100)

URSA MINOR
(Constellation of the Lesser Bear)

CAT.	LONG.	LAT.	MAG.	NAME	DESCRIPTION
1α	3♊50	+65;50	1.99	Polaris	The star on the tip of the tail
23δ	6♊26	+69;42	4.35	Yildun	The next star in the tail
22ε	14♊14	+73;39	4.22		The next star, before the beginning of the tail
16ζ	2♋10	+74;52	4.31		The southern one on the western side of the rectangle
21η	5♋26	+77;41	4.95		The northern one on the same side
7β	17♋58	+72;47	2.08	Kochab	The southern one of those on the eastern side
13γ	26♋04	+75;04	3.04	Pherkad	The northern one on the same side
5A	13♋10	+71;14	4.30		The unfigured star near it*

* <The star lying on a straight line with the stars in the rear side [of the rectangle] and south of them> (Toomer [1984], p 341)

URSA MAJOR
(Constellation of the Greater Bear)

CAT.	LONG.	LAT.	MAG.	NAME	DESCRIPTION
1o	28♊10	+40;04	3.37	Muscida	The star at the tip of the muzzle
2A	26♊46	+44;22	5.46		The western star of those in the two eyes
4π²	27♊59	+43;46	4.60		The eastern one of these
8ρ	29♊06	+47;41	4.75		The western star of the two in the forehead
13σ²	0♋23	+47;38	4.79		The eastern one of these
24d	1♋28	+50;58	4.58		The star at the end of the western ear
14τ	2♋40	+44;21	4.66		The western star of the two in the neck
23h	5♋54	+44;54	3.70		The eastern one of these
29υ	11♋24	+42;33	3.80		The northern one of the two in the breast
30φ	14♋28	+38;02	4.58		The southern one of these
25θ	12♋37	+35;06	3.18	Al Haud	The star in the left knee

CAT.	LONG.	LAT.	MAG.	NAME	DESCRIPTION
9τ	8♋06	+29;31	3.14	Talitha Borealis	The northern star <of the [two] in the front left paw>
12κ	9♋06	+28;47	3.58	Talitha Australis	The southern one of these
18e	8♋26	+35;50	4.81		The star above the right knee
15f	8♋18	+33;15	4.48		The star below the right knee
50α	20♋13	+49;32	1.80	Dubhe	Of those in the quadrilateral, the star on the back
48β	24♋27	+44;55	2.36	Merak or Mirak	Of these, the star on the flank
69δ	5♌58	+51;30	3.31	Megrez	The star at the beginning of the tail
64γ	5♌25	+46;59	2.43	Phecda or Phad	The remaining star in the left <hind> thigh
33λ	24♋44	+29;45	3.45	Tania Borealis	The western star of <the [two] in the left hind paw>
34μ	26♋24	+28;49	3.16	Tania Australis	The star east of this one
52ψ	3♌54	+35;25	3.00		The star in the left <knee>
54ν	11♌47	+26;03	3.50	Alula Borealis	The northern star of <the [two] in the right hand paw>
53ξ	12♌31	+25;01	3.78	Alula Australis	The southern one of these
77ε	13♌46	+54;12	1.76	Alioth	The first star of the three in the tail after the beginning
79ζ	20♌28	+56;17	2.06	Mizar	The middle one of these
85η	1♍50	+54;24	1.85	Alkaid or Benatnash	The third one at the end of the tail

Of the Unfigured Stars beneath the Greater Bear

α⌒	29♌43	+40;07	2.89	Cor Caroli	The star under the tail far to the south
β⌒	23♌11	+40;30	4.26	Chara	The dimmer star west of it
α†	17♋08	+17;47	3.14		The more southern star of those between the <front legs> of the Bear and the head of the Lion
38†	15♋44	+19.58	3.82		The star north of this
10Δ	18♋59	+20;33	4.58		The star east of the other three dim ones
BSC 3809*	17♋57	+23;37	4.81		The star west of this last
BSC 3612‡	12♋44	+20;41	4.56		The star still further west than this last
31†	2♋44	+22;55	4.25		The star between the <front legs> of the Bear and the Twins

⌒ Canum Venaticorum; † Lyncis; Δ Leonis Minoris
*The identification of this star [Peters and Knobel (Piazzi: IX 115)] is highly uncertain
‡ Peters and Knobel (Piazzi: VIII 2245)

APPENDIX I 153

DRACO
(Constellation of the Dragon)

CAT.	LONG.	LAT.	MAG.	NAME	DESCRIPTION
21μ	29♎34	+76;26	5.80	Arrakis	The star on the tongue
24ν¹	14♏55	+78;22	4.89		
25ν²	14♏59	+78;21	4.86	Kuma	The star in the mouth
23β	16♏54	+75;32	2.79	Rastaban	The star above the eye
32ξ	29♏41	+80;31	3.74	Grumium	The star in the jaw
33γ	3♐11	+75;13	2.23	Eltanin	The star above the head
39b	28♐09	+82;00	4.98		The northern star of the three in a straight line in the first <bend> of the neck
46c	5♉45	+78;06	5.04		The southern one of these
45d	1♉36	+80;01	4.79		The middle star of these
47o	21♉12	+81;00	4.65		The star east of this one
58π	10♓22	+81;49	4.58		The southern star of the western side of the square in the next <bend>
57δ	24♓05	+82;49	3.07	Al Tais	The northern star of the western side
63ε	9♈02	+79;21	3.84	Tyl	The northern star of the eastern side
67ρ	26♓42	+78;04	4.51		The southern star of the eastern side
61σ	13♈17	+80;39	4.67	Alsafi	The southern star of the triangle in the next <bend>
52υ	26♈52	+83;02	4.83		The western star of the remaining two of the triangle
60τ	0♉49	+80;29	4.45		The eastern star of these
31ψ¹	18♊01	+83;46	4.59	Dsiban	The eastern star of those in the next triangle west of this
44χ	23♉00	+83;08	3.57		The southern star of the remaining two of the triangle
43ø	17♉10	+84;38	4.21		The northern star of the remaining two
27f	28♋54	+86;45	5.05		The eastern star of the two small stars west of the triangle
28ω	16♋39	+86;49	4.79		The western one of these
18g	7♍43	+81;39	4.86		The <southernmost> of the next three in a straight line
19h	7♍54	+83;17	4.89		The middle one of the three
22ζ	5♍44	+84;46	3.17	Al Dhibah	The northern one of these
14η	18♍36	+78;29	2.72	Aldnibain	The northern one of the next two to the west
13θ	21♍44	+74;27	4.00		The southern one of these
12τ	9♍24	+71;07	3.29	Edasich	The western star of those to the west in the <bend> of the tail

CAT.	LONG.	LAT.	MAG.	NAME	DESCRIPTION
10i	9♌37	+65;15	4.65		The western star of the two rather distant from the last one
11α	12♌09	+66;16	3.65	Thuban	The eastern star of these
5κ	21♋09	+61;35	3.86		The star near them by the tail
1λ	15♋18	+57;02	3.84	Giauzar Giansar	The remaining star at the end of the tail

CEPHEUS
(Constellation of the Crowned King)

CAT.	LONG.	LAT.	MAG.	NAME	DESCRIPTION
1κ	8♉45	+75;15	4.38		The star <on the right leg>
35γ	5♉26	+64;23	3.20	Er Rai	The star <on the left leg>
8β	11♈26	+70;59	3.19	Alfirk	The star under the belt on the right side
5α	18♓36	+68;52	2.45	Alderamin	The star touching the right shoulder from above
3η	9♓26	+71;31	3.42		The star touching the right elbow from above
2θ	11♓04	+73;55	4.22		The star touching the same elbow from below
17ξ	29♓52	+65;39	4.28	Al Kurhah	The star in the chest
32ι	9♈05	+62;29	3.51		The star in the left arm
23ε	18♓31	+59;57	4.19		The southern star of the three in the tiara
21ζ	19♓41	+61;05	3.40		The middle one of the three
22λ	21♓43	+61;49	5.05		The northern one of the three

Of the Uncentered Stars about Cepheus

CAT.	LONG.	LAT.	MAG.	NAME	DESCRIPTION
13μ	15♓31	+64;09	4.54	Garnet Star	The star west of the tiara
27δ	23♓18	+59;28	4.30		The star east of the tiara

BOÖTES
(Constellation of the Ploughman)

CAT.	LONG.	LAT.	MAG.	NAME	DESCRIPTION
17κ²	4♍42	+58;53	4.52	Asellus Tertius	The <most advanced> [west] star of the three in the left hand
21ι	6♍02	+58;50	4.80	Asellus Secundus	The middle and southern one of the three
23θ	7♍11	+60;21	4.05	Asell Primus	The eastern one of the three
19λ	12♍01	+54;39	4.18		The star in the left elbow
27γ	22♍44	+49;35	3.04	Seginus	The star in the left shoulder
42β	29♍09	+54;17	3.51	Nekkar	The star in the head

CAT.	LONG.	LAT.	MAG.	NAME	DESCRIPTION
49δ	8♎04	+49;10	3.46		The star in the right shoulder
51μ¹	8♎15	+53;33	4.30	Alkalurops	The star north of these and <on the staff>
52ν¹	7♎24	+57;14	5.02		The star north of this at the tip of the <staff>
53ν²	7♎38	+57;22	5.00		
2η*	11♎57	+47;02	6.08		The northern one of the two below the shoulder in the cudgel
10*	11♎38	+46;08	5.51		The southern one of these
45c	10♎12	+40;41	4.93		The star at the <end> of the right <arm>
43ψ	8♎39	+42;22	4.53		The western one of the two in the wrist
46b	9♎57	+42;04	5.67		The eastern one of these
41ω	8♎51	+40;21	4.81		The star at the end of the <handle of the staff>
36ε	3♎10	+40;46	2.37	Izar or Mizar	The star in the right thigh in the girdle
28σ	28♍50	+42;09	4.46		The eastern one of the two in the girdle
25ρ	27♍54	+42;31	3.58		The western one of these
30ζ	8♎07	+28;02	4.43		The star in the right heel
8η	24♍23	+28;22	2.68	Muphrid or Mufrid	The northern one of the three in the left <lower leg>
4τ	23♍19	+26;41	4.49		The middle one of the three
5υ	24♍23	+25;17	4.04		The southern one of these
16α	29♍26	+32;09	−0.05	Arcturus	<The star between the thighs called 'Arcturus', reddish>

* Coronae Bor.

CORONA BOREALIS
(Constellation of the Northern Crown)

CAT.	LONG.	LAT.	MAG.	NAME	DESCRIPTION
5α	17♎14	+44;32	2.22	Alphecca	The bright star in the crown
3β	14♎17	+46;13	3.67	Nusakan	The most western of all
4θ	14♎27	+48;44	4.14		The one east and north of this one
9π	17♎12	+50;40	5.56		The one again east and north of this last
8γ	19♎59	+44;42	3.83		The star east of this bright one southwards
10δ	22♎07	+45;01	4.62		The star east of this last and nearby
13ε	24♎12	+46;19	4.14		The star still east of these
14ι	24♎04	+49;23	4.98		The star east of all those in the crown

HERCULES
(Constellation of the Kneeling Man)

CAT.	LONG.	LAT.	MAG.	NAME	DESCRIPTION
64α	21♏21	+37;32	3.58	Rasal-gethi	The star in the head
27β	6♏16	+42;58	2.78	Korne-phoros	The star in the right shoulder beside the armpit
20γ	4♏22	+40;14	3.80		The star in the right arm
7κ[a]	0♏52	+37;27	5.00	Maasim	The star in the right elbow
65δ	19♏55	+48;03	3.12	Sarin	The star in the left shoulder
76λ	25♏04	+49;33	4.40		The star in the left arm
86μ	0♐40	+51;50	3.41		The star in the left elbow
103o	7♐57	+52;27	3.83		The eastern one of the three in the left wrist
94ν	4♐42	+53;54	4.41		The northern one of the remaining two
92ξ	4♐22	+52;58	3.69	Marfik	The southern one of these
40ζ	6♏59	+53;11	2.80		The star in the right side
58ε	13♏28	+53;30	3.91		The star in the left side
59	13♏05	+56;09	5.21		The one north of this last, in the left buttock
61c	13♏08	+58;04	6.33		The star at the beginning of the same thigh
67π	17♏11	+59;49	3.16		The western star of the three in the left thigh
69e	18♏05	+60;21	4.64		The one east of this last
75ρ[a]	20♏31	+60;24	4.17		The star again east of this one
91θ	3♐42	+60;58	3.86		The star in the left knee
85ι	25♏00	+69;32	3.80		The star in the left shin
74x	15♏38	+69;15	5.58		The western one of the three in the foot
77y	17♏36	+71;28	5.84		The middle one of the three
82z	22♏36	+72;00	5.36		The eastern one of these
44η	3♏43	+60;34	3.49		The star at the beginning of the right thigh
35σ	28♎11	+63;22	4.19		The one north of this and in the same thigh
22τ	19♎13	+66;00	3.89		The star in the right knee
11φ	16♎29	+63;57	4.21		The southern star of the two under the right knee
6υ	13♎05	+64;30	4.73		The northern one of these
1χ	13♎04	+60;03	4.62		The star in the right shank
52ν¹*	7♎24	+57;14	5.02		The star at the end of the right <leg> which is the same as that at the tip of the <staff>
53ν²*	7♎38	+57;22	5.00		
24ω	6♏41	+35;27	4.58	Cujam	The star south of that in the right <upper> arm

* Boötis

APPENDIX I 157

LYRA
(Constellation of the Lyre)

CAT.	LONG.	LAT.	MAG.	NAME	DESCRIPTION
3α	20♐27	+61;51	0.03	Vega	The bright star on the shell called the Lyre
4ε¹	24♐01	+62;37	4.67		The northern one of the two lying next to it
5ε²	24♐01	+62;34	4.59		
6ζ¹	23♐28	+60;36	4.35		The southern one of these
7ζ²	23♐29	+60;35	5.73		
11δ¹	26♐48	+59;41	5.60		The star east of these and at the beginning of the horns <[of the lyre]>
12δ²	27♐05	+59;34	4.28		
20η	5♉33	+60;55	4.40	Aladfar	The northern one of the two near the eastern side of the shell
21θ	6♉00	+59;48	4.36		The southern one of these
10β	24♐15	+56;15	3.80	Sheliak	The northern one of the two western stars in the crossbar
9ν²	24♐00	+55;27	5.24		The southern one of these
14γ	27♐19	+55;16	3.24	Sulafat	The northern one of the two in the crossbar
15λ	27♐32	+54;41	4.93		The southern one of these

CYGNUS
(Constellation of the Bird or Swan)

CAT.	LONG.	LAT.	MAG.	NAME	DESCRIPTION
6β¹	6♉39	+49;12	3.08	Albireo	The star in the beak
12φ	10♉21	+50;50	4.69		The one east of this and in the head
21θ	18♉28	+54;29	3.89		The star in the middle of the <neck>
37γ	0♒26	+57;17	2.20	Sadr	The star in the breast
50α	11♒00	+60;01	1.24	Deneb	The bright star in the tail
18δ	21♉51	+64;36	2.86		The star in the <bend> of the right wing
13θ	24♉15	+69;40	4.48		The southern star of the three in the right wing spread
10ι²	23♉42	+71;34	3.78		The middle one of the three
1κ	20♉39	+73;57	3.79		The northern one at the edge of the wing spread
53ε	2♒54	+49;28	2.50	Gienah	The star in the <bend> of the left wing
54λ	5♒18	+51;46	4.53		The star north of <this> and in the middle of the same wing

CAT.	LONG.	LAT.	MAG.	NAME	DESCRIPTION
64ζ	8♒32	+43;51	3.20		The star at the edge of the left wing spread
58υ	11♒45	+55;02	3.93		The star in the left <leg>
62ξ	16♒26	+56;40	3.70		The star in the left knee
30o¹	3♒45	+63;50	4.82		The western one of the two in the right <leg>
32o²	5♒32	+64;25	3.94		The eastern one of these
45ω¹	11♒46	+64;09	4.94	Ruchba	<The nebulous star on the right knee>
46ω²	12♒32	+64;17	5.46		
65τ	13♒52	+50;28	3.74		The southern star of the two under the left wing
67σ	15♒55	+51;35	4.22		The northern one of these

CASSIOPEIA
(Constellation of the Enthroned Queen)

CAT.	LONG.	LAT.	MAG.	NAME	DESCRIPTION
17ζ	10♈32	+44;34	3.67		The star in the head
18α	13♈15	+46;29	2.22	Schedar	The star in the breast
24η	15♈34	+47;16	3.45	Achird	The star north of this and in the girdle
27γ	19♈24	+48;38	2.30	Tsih	The star above the <throne, just over the thighs>
37δ	23♈17	+46;16	2.67	Ruchbah	The star in the knees
45ε	0♉11	+47;20	3.36	Segin	The star in the <lower leg>
35ι	7♉37	+48;44	4.51		<The star on the end of the leg>
33θ	17♈09	+42;59	4.33	Al Marfik	The star in the left arm
34φ	20♈57	+44;54	4.98		The star below the left elbow
8σ	5♈38	+49;16	4.88		The star in the right forearm
15κ	18♈06	+52;06	4.16		The star above the foot of the throne
11β	10♈33	+51;14	2.26	Caph	The star in the middle of the back of the <throne>
7ρ	6♈35	+51;02	5.50		<The star on the top of the throne-back>

PERSEUS
(Constellation of the Hero or Champion)

CAT.	LONG.	LAT.	MAG.	NAME	DESCRIPTION
7*	29♈38	+40;31	5.99		The nebula in the right hand
15η	4♉03	+37;15	3.76	Miram	The star in the right elbow
23γ	5♉21	+34;17	2.93		The star in the right shoulder
13θ	29♈53	+31;29	4.11		The star in the left shoulder

APPENDIX I

CAT.	LONG.	LAT.	MAG.	NAME	DESCRIPTION
18τ	3♉15	+34;08	3.93		The star in the head
19ι	4♉05	+30;36	4.04		The star in the broad of the back
33α	7♉23	+29;54	1.79	Mirfak	The bright star in the right side
35σ	7♉55	+27;47	4.35		The western one of the three after the one in the side
37φ	9♉03	+27;45	4.22		The middle one of the three
39δ	10♉06	+27;04	3.01		The eastern one of these
27κ	2♉57	+25;57	3.80		The star in the left elbow
26β	1♉28	+22;12	2.31	Algol	The bright star in the Gorgon's head
28ω	1♉40	+20;44	4.61		The star east of this
25ρ	0♉10	+20;25	3.77		The star west of the bright one
22π	29♉12	+21;32	4.68		The star left farther west than this
7²b	17♉05	+28;13	4.60		The star in the right knee
47λ	15♉03	+28;39	4.28		The star west of this and above the knee
48c	14♉47	+26;00	4.03		The western star of the two above the <bend in the knee>
51μ	16♉05	+26;27	4.15		The eastern one, <just over the bend in the knee>
53d	16♉53	+24;23	4.84		The star in the right calf
58e	18♉51	+18;45	4.25		The star in the right ankle
41ν	9♉07	+21;54	3.77		The star in the left thigh
45ε	10♉57	+18;53	2.88		The star in the left knee
46ξ	10♉15	+14;41	4.04	Menkib	The star in the left <lower leg>
38ο	6♉25	+11;57	3.83	Atik	The star in the left heel
44ζ	8♉23	+11;05	2.89		The star east of it in the left foot

The Unfigured Stars about Perseus

CAT.	LONG.	LAT.	MAG.	NAME	DESCRIPTION
52f	14♉25	+18;41	4.69		The star east of the one in the <left> knee
BSC** 1314	17♉37	+31;29	5.19		The star north of <the one> in the right knee
16	27♈04	+20;49	4.22		The western star of those in the Gorgon's head

*7 Persei lies exactly at the location specified by Ptolemy. However, Toomer (1984), p 352 places here the famous double star cluster formed by Galactic Clusters 869 and 884
**BSC 1314 = H 14, where H = Hevelius catalogue number

AURIGA
(Constellation of the Charioteer)

CAT.	LONG.	LAT.	MAG.	NAME	DESCRIPTION
33δ	5♊08	+30;38	3.72		The southern one of the two at the head
30ξ	4♊23	+31;59	4.96		The northern one and above the head
13α	27♉06	+22;50	0.08	Capella	The star in the left shoulder called Cappella
34β	5♊11	+21;14	1.90	Menkalinan	The star in the right shoulder
32ν	3♊32	+15;26	3.98		The star in the right elbow
37θ	5♊10	+13;32	2.64		The star in the right wrist
7ε	24♉06	+20;40	3.06	Al Ma'az	The star in the left elbow
10η	24♉42	+18;03	3.17	Haedus II	The eastern one of the two in the left wrist called the Kids
8ζ	23♉53	+17;57	3.74	Haedus I	The western one of these
3ι	21♉53	+10;12	2.69	Hassaleh	The star in the left ankle
23γ*	27♉49	+5;13	1.65	Al Ka'b dhi'l Inan	The star common to the right ankle and the horn of the Bull
25χ	29♉25	+8;37	4.76		The star north of this <in the lower hem [of the garment]>
24φ	28♉28	+10;59	5.07		The star still north of this in the buttock
14	25♉46	+9;20	5.02		The little star above the left foot

(=112β Tauri). This bright star in Taurus is called El Nath ("the Butting One"*), which is the name it is usually known by, i.e. marking the tip of the upper horn of the Bull, rather than Al Ka'b dhi'l Inan (*"the Heel of the Rein-Holder"*), i.e. marking the heel of Auriga

OPHIUCHUS
(Constellation of Serpentarius, the Serpent Holder)

CAT.	LONG.	LAT.	MAG.	NAME	DESCRIPTION
55α	27♏34	+36;14	2.08	Rasalhague	The star in the head
60β	0♐36	+28;08	2.77	Celbalrai or Cheleb	The western one of the two in the right shoulder
62γ	1♐53	+26;26	3.74	Muliphen	The eastern one of these
25ι	15♏51	+32;47	4.37		The western star of the two in the left shoulder
27κ	17♏12	+32;08	3.39		The eastern one of these
10λ	10♏48	+23;52	3.81	Marfik	The star in the left elbow
1δ	7♏33	+17;25	2.73	Yed Prior	The western star of the two in the left hand
2ε	8♏39	+16;42	3.23	Yed Post.	The eastern one of these
57μ	29♏33	+15;29	4.62		The star in the right elbow
64ν	5♏00	+13;53	3.33		The western star of the two in the right hand

APPENDIX I

CAT.	LONG.	LAT.	MAG.	NAME	DESCRIPTION
69τ	6♐01	+15;31	4.77		The eastern one of these
35η	23♏11	+7;32	2.43	Sabik	The star in the right knee
40ξ	26♏02	+2;10	4.39		The star in the right <lower leg>
36A	25♏34	−3;55	4.32		The western star of the four in the right foot
42θ	26♏38	−1;35	3.26		The one east of this one
44b	27♏35	−0;45	4.16		The star still east of this last
51c	28♏43	−0;25	4.80		The remaining and easternmost of the four
52	29♏31	+1;33	6.47		The one east of these and touching the heel
13ζ	14♏26	+11;40	2.57		The star in the left knee
8φ	13♏56	+5;26	4.27		The northern one of the three in a straight line in the left <lower leg>
7χ	13♏13	+3;28	4.59		The middle one of these
4ψ	12♏48	+1;47	4.48		The southern one of the three
9ω	14♏51	+0;43	4.44		The star in the left heel
5ρ^a	13♏41	−1;31	4.56		The star touching the hollow of the left foot

The Unfigured Stars about the Serpentarius

CAT.	LONG.	LAT.	MAG.	NAME	DESCRIPTION
66n	5♐19	+28;06	4.70		The northern one of the three east of the right shoulder
67	5♐25	+26;39	3.96		The middle one of the three
68	5♐43	+25;02	4.42		The southern one of these
70	6♐36	+26;51	4.02		The star east of the three, above the middle one
72	7♐27	+33;13	3.73		The lone star north of the four

SERPENS
(Constellation of the Serpent of Serpentarius)

CAT.	LONG.	LAT.	MAG.	NAME	DESCRIPTION
21ι	22♎18	+38;20	4.51		Of the square in the head, the star at the end of the jaw
38ρ	24♎39	+40;13	4.76		The star touching the nostrils
41γ	27♎28	+36;06	3.84		The star in the temple
28β	25♎01	+34;33	3.66	Chow	The star <where the neck joins [the head]>
35κ	24♎54	+37;23	4.09		The star in the middle of the square <in the mouth>
44π	27♎13	+42;39	4.82		The star outside and north of the head
13δ	23♎32	+29;06	3.80		The star after the first curve of the neck

CAT.	LONG.	LAT.	MAG.	NAME	DESCRIPTION
27λ	27♎41	+26;50	4.42		The northern one of the three following this
24α	27♎09	+25;41	2.60	Unukalhai	The middle one of the three
37ε	29♎26	+24;11	3.70		The southern one of these
32μ	1♏11	+16;28	3.54		The star after the next curve, west of the Serpentarius' left hand
3υ*	11♏49	+13;29	4.62		The star east of those in the hand
53ν	25♏30	+10;32	4.31		The star after the <back of the> Serpentarius' right thigh
55ξ	29♏48	+8;11	3.53		The southern one of the two east of this
56o	0♐40	+10;44	4.23		The northern one of these
57ζ	5♐16	+20;00	4.62		The star after the right hand in the tail's curve
58η	11♐15	+20;17	3.25		The star east of this likewise in the tail
63θ¹	21♐00	+27;08	4.62	Alya	The star at the tip of the tail

*Ophiuchi

SAGITTA
(Constellation of the Arrow)

CAT.	LONG.	LAT.	MAG.	NAME	DESCRIPTION
12γ	12♉22	+39;24	3.48		The lone star <on the arrow head>
8ζ	9♉23	+39;38	5.01		The eastern one of the three in the shaft
7δ	8♉45	+39;08	3.78		The middle one of these
5α	6♉25	+39;03	4.37		The western one of the three
6β	6♉34	+38;29	4.37		The star at the extremity of the notched end

AQUILA
(Constellation of the Eagle)

CAT.	LONG.	LAT.	MAG.	NAME	DESCRIPTION
63τ	10♉21	+27;15	5.51		The star in the middle of the head
60β	7♉47	+27;10	3.71	Alshain	The star west of this and in the neck
53α	6♉42	+29;23	0.76	Altair	The bright one <on the place between the shoulders>, called the Eagle
59ξ*	7♉54	+29;01	4.70		The one near this to the north
50γ	6♉15	+31;29	2.71	Tarazed	The western one of the two in the left shoulder
61φ	9♉15	+31;44	5.29		The eastern one of these
38μ	2♉00	+29;01	4.45		The western one of the two in the right shoulder
44σ	3♉07	+26;43	5.18		The eastern one of these

APPENDIX I   163

CAT.	LONG.	LAT.	MAG.	NAME	DESCRIPTION
17ζ	25♐07	+36;30	2.98	Deneb al Okab	The star farther off under the Eagle's tail touching the Milky Way

The Unfigured Stars about the Eagle

CAT.	LONG.	LAT.	MAG.	NAME	DESCRIPTION
55η	5♉43	+21;46	4.30		The western star of the two south of the Eagle's head
65θ	10♉11	+18;58	3.23		The eastern one of these
30δ	28♐45	+25;03	3.36		The star southwest of the Eagle's right shoulder
41ι	1♉07	+20;15	4.35		The star south of this
39κ	0♉07	+14;36	4.95		The star still south of this
16λ	22♐36	+17;47	3.43	Al Thalimain	The star west of all these

*Identification uncertain. Toomer (1984), p 357, identifies this star with o Aquilae

DELPHINUS
(Constellation of the Dolphin)

CAT.	LONG.	LAT.	MAG.	NAME	DESCRIPTION
2ε	19♉24	+29;17	4.03	Al Dhanab al Dulfim	The western star of the three in the tail
5ι	20♉39	+29;02	5.42		The northern one of the remaining two
7κ	20♉23	+27;44	5.07		The southern one of these
6β	21♉39	+32;09	3.64	Rotanev	The southern one of the western side of those stars in the rhomboidal figure of four sides
9α	22♉43	+33;13	3.77	Sualocin	The northern one of the western side
11δ	23♉30	+32;09	4.40		The southern one of the eastern side of the rhombus
12γ¹	24♉48	+32;59	5.14		The northern one of the eastern side
12γ²	24♉49	+32;59	4.27		
3η	20♉08	+30;52	5.39		The southern one of the three between the tail and the rhombus
4ζ	21♉05	+32;21	4.68		The western one of the remaining two northern ones
8θ	21♉35	+30;48	5.69		The remaining eastern one of these

EQUULEUS
(Constellation of the Foal)

CAT.	LONG.	LAT.	MAG.	NAME	DESCRIPTION
8α	28♑25	+20;21	3.92	Kitalpha	The western one of the two in the head
10β	0♒44	+21;11	5.15		The eastern one of these
5γ	28♑46	+25;27	4.69		The western one of the two in the <mouth>
7δ	29♑50	+25;04	4.47		The eastern one of these

PEGASUS
(Constellation of the Winged Horse)

CAT.	LONG.	LAT.	MAG.	NAME	DESCRIPTION
δ*	19♓39	+25;44	2.07	Alpheratz or Sirrah	The star common to the Horse's navel and Andromeda's head
88γ	14♓28	+12;34	2.83	Algenib	The star in the loin and at the end of the wing
53β	4♓38	+31;07	2.52	Scheat	The star in the right shoulder and at the beginning of the <leg>
54α	28♒48	+19;28	2.48	Markab	The star in the broad of the back and shoulder of the wing
62τ	6♓25	+25;35	4.59	Salm	The northern one of the two in the body under the wing
68υ	7♓14	+24;49	4.40		The southern one of these
44η	1♓09	+35;09	2.94	Matar	The northern one of the two in the right knee
43ο	0♓21	+34;27	4.80		The southern one of these
47λ	28♒25	+28;51	3.94	Sad al Nazi	The western one of the two close together in the chest
48μ	29♒43	+29;28	3.49	Sad al Bari	The eastern one of these
42ζ	21♒26	+17;46	3.40	Homam	The western one of the two close together in the neck
46ξ	23♒16	+18;47	4.19		The eastern one of these
50ρ	23♒50	+14;33	4.90		The southern one of the two in the mane
49σ	23♒21	+15;51	5.17		The northern one of these
26θ	11♒59	+16;30	3.51	Baham or Al Hawaim	The northern one of the two close together in the head
22ν	10♒32	+15;47	4.84		The southern one of these
8ε	7♒12	+22;14	2.10	Enif	The star in the muzzle
29π²	25♒03	+41;02	4.29		The star in the right ankle
24ι	19♒39	+34;23	3.76		The star in the left knee
10κ	14♒21	+36;44	4.14		The star in the left ankle

* = α Andromedae

ANDROMEDA
(Constellation of the Chained Woman)

CAT.	LONG.	LAT.	MAG.	NAME	DESCRIPTION
31δ	27 ♓ 08	+24;19	3.27		The star in the broad of the back
29π	28 ♓ 03	+27;03	4.34		The star in the right shoulder
30ε	26 ♓ 27	+23;00	4.36		The star in the left shoulder
25σ	25 ♓ 51	+31;30	4.51		The southern star of the three in the right <upper> arm
24θ	26 ♓ 39	+33;17	4.61		The northern one of these
27ρ	27 ♓ 01	+32;19	5.17		The middle one of the three
17ι	21 ♓ 33	+40;57	4.28		The southern star of the three <on the right hand>
19κ	22 ♓ 45	+41;40	4.13		The middle one of these
16λ	23 ♓ 54	+43;56	3.83		The northern one of the three
34ζ	25 ♓ 58	+17;33	4.03		The star in the left <upper> arm
38η	27 ♓ 43	+15;50	4.41		The star in the left elbow
43β	5 ♈ 42	+25;54	2.04	Mirach	The southern one of the three above the girdle
37μ	4 ♈ 28	+29;32	3.86		The middle one of these
35ν	4 ♈ 32	+32;26	4.53		The northern one of the three
57γ¹	19 ♈ 34	+27;39	2.26	Almach	The star above the left foot
54*	19 ♈ 59	+36;40	4.06		The star in the right foot
51†	17 ♈ 50	+35;18	3.57		The star south of this one
50υ	14 ♈ 06	+28;59	4.09		The northern one of the two in the left bend of the knee
53τ	14 ♈ 15	+27;46	4.95		The southern one of these
42φ	11 ♈ 49	+36;12	4.26		The star in the right knee
49A	15 ♈ 31	+34;24	5.27		The northern one of the two in the train (lower hem)
52χ	15 ♈ 52	31;18	5.00		The southern one of these
10	13 ♓ 16	+43;43	3.62		The star outside and west of the three in the right hand

* = φ Persei; † also known as υ Persei lies within the modern constellation of Andromeda

α Andromedae, called Alpheratz or Sirrah ('the Horse's Navel'), is not listed under Andromeda since it is listed by Ptolemy above as δ Pegasi

TRIANGULUM
(Constellation of the Triangle)

CAT.	LONG.	LAT.	MAG.	NAME	DESCRIPTION
2α	12 ♈ 13	+16;46	3.41	Ras al Muthallath	The star at the <apex> of the Triangle
4β	17 ♈ 36	+20;26	3.00		The western one of the three in the base

166 HISTORY OF THE ZODIAC

CAT.	LONG.	LAT.	MAG.	NAME	DESCRIPTION
8δ	18♈25	+19;27	4.86		The middle one of these
9γ	18♈49	+18;48	4.00		The eastern one of the three

CONSTELLATIONS OF THE STARS IN THE ZODIAC

ARIES
(Constellation of the Ram)

CAT.	LONG.	LAT.	MAG.	NAME	DESCRIPTION
5γ¹	8♈26	+7;05	4.83	Mesartim	The western star of the two in the horn
5γ²	8♈26	+7;05	3.88		
6β	9♈13	+8;25	2.64	Sharatan or Sheratan	The eastern one of these
17η	13♈19	+7;16	5.24		The northern one of the two in the muzzle
22θ	14♈08	+5;35	5.59		The southern one of these
8ι	8♈46	+5;20	5.10		The star in the neck
32ν	19♈24	+5;59	5.34		The star in the loins
48ε	23♈46	+3;58	4.63		The star at the beginning of the tail
57δ	26♈01	+1;38	4.34	Botein	The western one of the three in the tail
58ζ	27♈13	+2;43	4.89		The middle one of the three
63τ²	28♈55	+1;54	5.09		The eastern one of these
45ρ²	22♈08	+1;20	5.81		\<The star in the back of the thigh\>
45ρ³	22♈04	+1;08	5.59		
43σ	20♈10	−1;28	5.50		The star under the bend of the knee
87μ*	17♈01	−5;41	4.27		The star in the hind foot

Unfigured Stars around the Ram

CAT.	LONG.	LAT.	MAG.	NAME	DESCRIPTION
13α	12♈52	+9;55	2.00	Hamal	The star above the head which Hipparchus placed in the muzzle
41c	23♈28	+10;19	3.61		The eastern and brightest one of the four above the loins
39	23♈36	+12;22	4.51		The northern one of the remaining fainter three
35	22♈13	+11;07	4.65		The middle one of the three
33	21♈24	+10;44	5.30		The southern one of these

*Ceti

APPENDIX I   167

TAURUS
(Constellation of the Bull)

CAT.	LONG.	LAT.	MAG.	NAME	DESCRIPTION
5f	28♈48	−6;08	4.11		The northern one of the four in the <cut-off> section
4s	28♈19	−7;39	5.14		<The one close by this>
2ξ	27♈06	−8;59	3.73		The one next to this last
10	26♈25	−9;31	3.60		The southernmost one of the four
30e	2♉34	−8;51	5.07		The one east of these in the right shoulder blade
35λ	5♉52	−8;12	3.45		The star in the chest
49μ	8♉47	−12;25	4.28		The star in the right knee
38ν	5♉07	−14;41	3.90		The star in the right ankle
90c¹	14♉55	−9;45	4.26		The star in the left knee
88d	13♉59	−11;58	4.24		The star in the left <lower leg>
54γ	10♉58	−5;58	3.63	Hyadum I	Of those in the face called the Hyades, the one on the nostrils
61δ¹	12♉03	−4;12	3.75	Hyadum II	The star between this last and the northern eye
61δ²	12♉18	−4;20	4.80		
77θ¹	13♉08	−5;59	3.84		The star between this last one and the southern eye
78θ²	13♉08	−6;04	3.40		
87α	15♉00	−5;37	0.86	Aldebaran	The bright red star of the Hyades in the southern eye
74ε	13♉39	−2;48	3.53	Ain	The other one in the northern eye
97i	18♉57	−3;53	5.10		The star at the beginning of the southern horn and the ear
104m	22♉28	−4;29	4.91		The southern one of the two in the southern horn
106l	23♉02	−2;43	5.27		The northern one of these
123ζ	0♊01	−2;28	3.02	Al Hecka	The star at the tip of the southern horn
94τ	17♉24	+0;27	4.27		The star at the beginning of the northern horn
112β	27♉49	+5;13	1.65	El Nath	The star at the tip of the northern horn, the same as the one in the Charioteer's right foot
69υ	13♉41	+0;53	4.28		The northern one of the two close together in the northern ear
65κ¹	13♉24	+0;24	4.21		The southern one of these
37A¹	8♉39	+1;03	4.36		The western one of the two small ones in the neck

CAT.	LONG.	LAT.	MAG.	NAME	DESCRIPTION
50ω²	11♉19	−0;58	4.91		The eastern one of these
44p	10♉56	+5;05	5.47		The southern one of the western side of the square in the neck
42ψ	10♉36	+7;41	5.27		The northern one of the western side
59χ	13♉21	+3;46	5.39		The southern one of the eastern side
52φ	13♉11	+5;36	4.95	Al Kalbain	The northern one of the eastern side
19e	4♉48	+4;19	4.29	Taygeta or Taygete	The northern limit of the <western> side of the Pleiades
23d	4♉57	+3;45	4.18	Merope	The southern limit of the <western> side
25η	5♉14	+3;50	2.87	Alcyone	The eastern and narrowest limit of the Pleiades
27f	5♉36	3;42	3.62	Atlas	
BSC 1188*	6♉11	+5;11	4.81		The small star outside and north of the Pleiades

Unfigured Stars around the Bull

CAT.	LONG.	LAT.	MAG.	DESCRIPTION
10	27♈21	−18;25	4.29	The star below the right foot and the shoulder blade
102ι	22♉00	−1;27	4.61	The western one of the three above the southern horn
109n	25♉49	−1;14	4.93	The middle one of the three
114o	27♉44	−1;34	4.87	The eastern one of these
126	0♊43	−7;06	4.84	The northern one of the two below the tip of the southern horn
129	2♊02	−7;51	6.01	The southern one of these
121	29♉38	+0;27	5.37	The western star of the five eastern ones under the northern horn
125	0♊40	+2;16	5.16	The one east of this
132	2♊45	+0;53	4.88	The one again east of this
136	3♊45	+3;55	4.58	The northern one of the two remaining eastern ones
139	4♊47	+2;14	4.82	The southern one of these

*BSC 1188 = Peter's and Knobel (Piazzi: III 170)

APPENDIX I 169

GEMINI
(Constellation of the Twins)

CAT.	LONG.	LAT.	MAG.	NAME	DESCRIPTION
66α	25♊34	+9;56	1.58	Castor	The star in the head of the western Twin
78β	28♊45	+6;31	1.16	Pollux	The red star in the head of the eastern Twin
34θ	16♊21	+10;47	3.59		<The star in the left forearm> of the western Twin
46τ	20♊41	+7;31	4.41		The star in the same arm
60ι	24♊15	+5;33	3.79		The star east of this one and in the broad of the back
69υ	26♊35	+5;01	4.05		The one east of this in the right shoulder of the same Twin
77κ	28♊55	+2;51	3.57		The star in the eastern shoulder of the eastern Twin
57A	24♊07	+2;43	5.03		The star in the right side of the western Twin
58	24♊24	+0;38	6.16		The star in the left side of the eastern Twin
27ε	15♊11	+1;48	3.00	Mebsuta	The star in the left knee of the western Twin
43ζ	20♊14	−2;18	4.16	Mekbuda	The star under the left knee of the eastern Twin
55δ	23♊46	−0;26	3.50	Wasat or Wesat	The star in the left <groin> of the eastern Twin
54λ	24♊03	−5;52	3.58		The star over the bend of the right knee of the same Twin
7η	8♊42	−1;09	3.65	Propus or Tejat Prior	The star in the forward foot of the western Twin
13μ	10♊31	−1;02	2.89	Nuhatai or Tejat Post.	The star east of this in the same foot
18ν	12♊03	−3;19	4.15		<The star on the right foot> of the western Twin
24γ	14♊19	−7;00	1.92	Alhena or Almeisan	<The star on the left foot> of the eastern Twin
31ξ	16♊31	−10;16	3.35	Al Zirr	The star in the right foot of the eastern Twin

Unfigured Stars around the Twins

CAT.	LONG.	LAT.	MAG.	NAME	DESCRIPTION
1H	6♊11	−0;24	4.15		The western star of the forward foot of the western Twin
44κ*	8♊38	+5;59	4.34		The bright western star of the western knee

CAT.	LONG.	LAT.	MAG.	NAME	DESCRIPTION
36d	17♊12	−1;24	5.24		The western star of the left knee of the eastern Twin
85	2♋18	−1;06	5.37		The northern star of the three in the straight line east of the right arm of the eastern Twin
81g	0♋22	−2;52	4.88		The middle one of the three
74f	28♊56	−3;59	5.05		The southern one of these between the forearm and hand
16ζ¹+	6♋33	−2;25	5.63	Tegmine	The bright one east of these three

*Aurigae; +Cancri—the name Tegmine means 'In the covering', relating to the covering part of the rear edge of the Crab's shell

CANCER
(Constellation of the Crab)

CAT.	LONG.	LAT.	MAG.	NAME	DESCRIPTION
41ε	12♋39	+0;58	6.30	Praesepe 'Beehive' (Messier 44)	The middle of the nebula called <Praesepe>, in the breast
33η	10♋40	+1;23	5.33		The northern star of the two western ones of the square about the nebula
31θ	11♋00	−0;57	5.33		The southern star of the two western ones
43γ	12♋49	+3;00	4.66	Asellus Borealis	The northern star of the two eastern ones of the square called the Asses <Aselli>
47δ	13♋56	−0;01	3.93	Asellus Aus.	The southern one of these two
65α	18♋53	−5;16	4.25	Acubens	The star in the southern claw
48ι[a]	11♋34	+10;14	4.02		The star in the northern claw
10μ²	4♋42	+1;09	5.29		The star in the northern <back leg>
17β	9♋33	−10;29	3.52	Al Tarf	The star in the southern <back leg>

Unfigured Stars around the Crab

CAT.	LONG.	LAT.	MAG.	NAME	DESCRIPTION
62o¹	17♋36	−2;04	5.19		The star above the joint of the southern claw
63o²	17♋36	−1;48	5.65		
76κ	21♋26	−5;45	5.23		The star east of the tip of the southern claw
69υ	16♋16	+7;05	5.45		The western star of the two above the nebula
77ξ	18♋26	+5;13	5.17		The eastern one of these

Uncertain identification of the unfigured stars. Toomer (1984), p 366 identifies o¹+ o² as π Cancri.

ζ Cancri, called Tegmine ('In the covering'), is not listed under Cancer since it is listed by Ptolemy above as an unfigured star belonging to Gemini

LEO
(Constellation of the Lion)

CAT.	LONG.	LAT.	MAG.	NAME	DESCRIPTION
1κ	20♋30	+10;16	4.46		The star at the tip of the nostrils
4λ	23♋05	+7;43	4.31	Alterf	The star in the open mouth
24μ	26♋43	+12;15	3.88	Rasalas	The northern one of the two in the head
17ε	25♋56	+9;33	2.98	Ras Elased	The southern one of these
36ζ	2♌45	+11;43	3.43	Aldha-fera or Adhafera	The northern one of the three in the <neck>
41γ¹	4♌37	+8;42	2.01	Algieba	The middle one of the three, nearby
41γ²	4♌37	+8;42	3.80		
30η	3♌08	+4;43	3.51		The southern one of these
32α	5♌12	+0;22	1.36	Regulus	The one at the heart called Regulus
31A	5♌42	−1;30	4.38		The one south of this, in the chest
27ν	2♌35	−0;05	5.25		The star a little west of the one in the heart
16ψ	28♋43	+0;11	5.35		The star in the right knee
5ξ	26♋56	−3;16	4.97		The star in the right foreclaw
14o	29♋34	−3.52	3.52		The star in the left foreclaw
29π	4♌35	−4;02	4.69		The star in the left knee
47ρ	11♌38	+0;02	3.87		The star in the left armpit
46i	9♌43	+4;28	5.44		The western star of the three in the belly
52k	12♌55	+5;53	5.50		The northern one of the other eastern two
53l	14♌54	+2;44	5.31		The southern one of these
60b	14♌03	+12;48	4.41		The western one of the two in the loin
68δ	16♌23	+14;17	2.56	Zosma	The eastern one of these
81†	20♌49	+11;41	5.61		The northern one of the two in the buttocks
70θ	18♌38	+9;39	3.33	Chort	The southern one of these
78ι	22♌40	+6;03	4.00		The star in the calves of the legs
77σ	23♌59	+1;40	4.04		The star in the hindleg joints
84τ	26♌45	−0;36	4.95		The star south of this one, in the <lower legs>
91υ	0♍17	−3;03	4.30		The star in the hindclaws
94β	27♌02	+12;24	2.13	Denebola	The star at the tip of the tail

Unfigured Stars around the Lion

CAT.	LONG.	LAT.	MAG.	NAME	DESCRIPTION
41LMi	8♌43	+13;52	5.07		The western star of the two above the back

172   HISTORY OF THE ZODIAC

CAT.	LONG.	LAT.	MAG.	NAME	DESCRIPTION
54	10♌42	+16;24	4.33		The eastern one of these
63χ	19♌55	+1;22	4.62		The northern one of the three <under> the flank
59c	19♌16	−0;16	4.99		The middle one of these
58d	20♌10	−2;35	4.84		The southern one of these
15γ*	29♌00	+28;26	4.35		The northernmost part of the nebula called Coma the Hair, lying between the extremities of the Lion and the Bear
7h*	28♌46	+23;27	4.96		The western one of the <southern outrunners of Coma> the Hair
23k*	3♍36	+24;07	4.79		The eastern one of these in the figure of the ivy leaf

†Toomer (1984), p367 suggests 81 Leonis but points out that this is extremely uncertain
*Comae Berenices

VIRGO
(Constellation of the Maiden)

CAT.	LONG.	LAT.	MAG.	NAME	DESCRIPTION
3ν	29♌21	+4;40	4.03		The southern star of the two in the <top> of the skull
2ξ	28♌31	+6;05	4.81		The northern one of these
9ο	3♍01	+8;32	4.11		The northern one of the two in the face east of these
8π	2♍46	+6;09	4.66		The southern one of these
5β	1♍56	+0;39	3.59	Zavijava	The star at the tip of the southern and left wing
15η	10♍06	+1;24	3.88	Zaniah	The western star of the four in the left wing
29γ	15♍39	+2;56	3.68	Porrima	The one east of this
46	20♍29	+2;58	5.98		The star east of this again
51θ	23♍30	+1;49	4.37		The last and eastern one of these four
43δ	16♍54	+8;48	3.38	Auva	The star in the right side under the girdle
30ρ	10♍37	+13;36	4.88		The western star of the three in the right and northern wing
32d²	12♍40	+11;37	5.21		The southern one of the two remaining ones
47ε	15♍15	+16;18	2.82	Vindemiatrix	The northern one of these called Vindemiatrix
67α	29♍07	−1;57	0.97	Spica	The star in the left hand called Spica
79ζ	27♍29	+8;50	3.37	Heze	The star under the girdle on the right buttock
74l²	28♍53	+3;14	4.68		The northern star of the western side of the square in the left thigh

CAT.	LONG.	LAT.	MAG.	NAME	DESCRIPTION
76h	0♎31	−0;19	5.20		The southern star of the western side
82m	1♎59	+1;53	5.00		The northern star of the two of the eastern side
68i	0♎05	−3;13	5.24		The southern one of the eastern side
86	4♎17	−1;15	5.50		The star in the left knee
90p	2♎26	+9;44	5.15		The star in the <back> of the right thigh
99ι	9♎06	+7;07	4.07	Syrma	The middle one of the three in the train about the <feet>
98κ	9♎41	+3;09	4.17		The southern one of these
105φ	10♎43	+11;57	4.81		The northern one of the three
100λ	12♎12	+0;41	4.52	Khambalia	The star in the left and southern foot
107μ	15♎20	+9;39	3.87	Rijl al 'Awwa	The star in the right and northern foot

Unfigured Stars around the Maiden

CAT.	LONG.	LAT.	MAG.	NAME	DESCRIPTION
26χ	17♍27	−3;24	4.64		The western star of the three in a straight line under the left forearm
40ψ	21♍28	−3;21	4.84		The middle one of these
49	24♍59	−3;10	5.18		The eastern one of the three
53	28♍04	−7;59	5.04		The western one of the three in a straight line under Spica
61	1♎05	−9;32	4.73		The middle one of these which is double
63	1♎07	−8;14	5.36		
89	7♎17	−6;14	4.97		The eastern one of these

61+63 Virginis is not a double star. This identification is extremely doubtful

LIBRA
(Constellation of the Balance or Scales)

CAT.	LONG.	LAT.	MAG.	NAME	DESCRIPTION
9α^1	20♎20	+0;32	5.15	Zuben–elgenubi or Kiffa Aus.	The bright star at the end of the southern claw (of the Scorpion)
9α^2	20♎23	+0;31	2.74		
7μ	19♎26	+2;14	5.31		The one north of and dimmer than this last
27β	24♎38	+8;43	2.61	Zubenescha-mali or Kiffa Bor.	The bright star of those at the end of the northern claw
19δ	20♎32	+8;27	5.06		The dim one west of this
24ι^1	26♎17	−1;39	4.54		The star in the middle of the southern claw
24ι^2	26♎30	−1;26	6.07		

CAT.	LONG.	LAT.	MAG.	NAME	DESCRIPTION
21ν	24♎02	+1;24	5.17		The star west of this in the same claw
38γ	0♏20	+4;37	3.91		The star in the middle of the northern claw
46θ	5♏02	+3;47	4.13		The one east of this in the same claw

Unfigured Stars around the Balance or Scales

CAT.	LONG.	LAT.	MAG.	NAME	DESCRIPTION
37	28♎46	+8;57	4.61		The western star of the three northern ones in the northern claw
48	5♏38	+6;19	4.85		The southern star of the two eastern ones
51*	6♏34	+9;28	4.16		The northern one of these
45λ	5♏43	+0;19	5.02		The eastern one of the three between the claws
43κ†	3♏02	+0;10	4.72		The northern one of the two remaining ones
42†	3♏35	−3;54	4.96		The southern one of these
20σ⁺	26♎00	−7;27	3.29	Zuban al Akrab or Zubenelakrab	The western star of the <3 stars south> of the southern claw
39υ	3♏52	−8;16	3.59		The northern one of the two remaining eastern ones
40τ	4♏38	−9;48	3.65		The southern one of these

* = ξ Scorpii;   ⁺ = γ Scorpii. The name Zubenelakrab means the Scorpion's Claw
† Peters and Knobel identify these two stars as κ and 0h. Arg 14782 = BSC 5810 — the latter is also adopted by Toomer, but in terms of stellar magnitude 42 Librae seems more likely

SCORPIUS
(Constellation of the Scorpion)

CAT.	LONG.	LAT.	MAG.	NAME	DESCRIPTION
8β¹	8♏26	+1;15	2.56	Graffias	The northern one of the three bright ones in the forehead
8β²	8♏26	+1;15	4.90		
7δ	7♏49	−1;45	2.31	Dschubba	The middle one of these
6π	8♏12	−5;14	2.90		The southern one of the three
5ρ	8♏25	−8;22	3.87		The star south of this again, in one of the <legs>
14ν	9♏53	+1;53	6.30	Jabbah	The northern one of the two lying beside the northernmost of the bright ones
9ω¹	8♏55	+0;28	3.95	Jabhat al Akrab	The southern one of these

APPENDIX I 175

CAT.	LONG.	LAT.	MAG.	NAME	DESCRIPTION
10ω¹	9♏04	+0;16	4.30		
20σ	13♏03	−3;47	2.90	Al Niyat	The western star of the three bright ones in the body
21α	15♏01	−4;19	1.02	Antares	The red middle one of these called Antares
23τ	16♏43	−5;52	2.82		The eastern one of the three
13c²	11♏30	−6;26	4.58		The western one of the two beneath <these, approximately on the last leg>
31d	12♏57	−6;55	4.78		The eastern one of these
26ε	20♏54	−11;35	2.29	Wei	The star in the first joint from the body
189μ¹	21♏25	−15;10	3.08		The star after this one in the second joint
193μ²	21♏31	−15;07	3.56		
198ζ¹	22♏23	−19;23	4.73		The northern one of the double star in the third joint
206ζ²	22♏34	−19;30	3.61		The southern one of the double star
302η	26♏00	−20;05	3.32		The next one in the fourth joint
138θ	0♐50	−19;22	1.85	Sargas	The star after this one in the fifth joint
210ι¹	2♐46	−16;26	3.02		The next to this last in the sixth joint
210ι²	3♐17	−16;24	4.80		
174κ	1♐43	−15;23	2.40	Girtab	The star in the seventh joint, <the joint next to the sting>
35λ	29♏50	−13;32	1.62	Shaula	The eastern one of the two in the <sting>
24υ	29♏16	−13;45	2.69	Lesath	The western one of these

Unfigured Stars around the Scorpion

CAT.	LONG.	LAT.	MAG.	NAME	DESCRIPTION
G Sco*	3♐58	−11;06	3.30		The nebula east of the <sting>
45d⁺	28♏07	−6;26	4.28		The western star of the two north of the <sting>
3‡	2♐29	−4;09	4.84		The eastern one of these

In contrast to Peters and Knobel, Toomer (1984), p 372 reverses ζ¹ and ζ², placing ζ² first
* = the globular cluster identified as G Scorpii (BSC 6630);   ⁺ Ophiuchi;   ‡ Sagittarii.
γ Scorpii, known as Zubenelakrab ('*the Scorpion's Claw*'), is not listed under Scorpio since it is listed by Ptolemy above as σ Librae

SAGITTARIUS
(Constellation of the Archer)

CAT.	LONG.	LAT.	MAG.	NAME	DESCRIPTION
10γ²	6♐32	−6;49	2.98	Al Nasl	The star at the tip of the arrow
19δ	9♐48	−6;13	2.70	Kaus Media	The star in the <[bow]> grip <held by the left hand>
20ε	10♐20	−10;51	1.83	Kaus Australis	The star in the southern part of the bow
22λ	11♐34	−1;58	2.82	Kaus Borealis	The southern star of those in the northern part of the bow
13μ	8♐27	+2;37	3.82	Polis	The northern of these, at the tip of the bow
15	8♐48	+2;57	5.34		
34σ	17♐37	−3;13	2.06	Nunki	The star in the left shoulder
27φ	15♐23	−3;41	3.17		The star west of this, <just over the arrow>
32ν¹	17♐42	+0;23	4.82	Ain al Rami	The nebular and double star in the eye
35ν²	17♐53	+0;26	4.99		
37ξ¹	18♐39	+2.23	5.06		The western star of the three in the head
37ξ²	18♐40	+1;56	3.51		
39ο	20♐11	+1;06	3.76		The middle one of these
41π	21♐29	+1;41	2.90	Al Baldah	The eastern one of the three
43d	23♐36	+3;31	4.90		The southern one of the three in the northern <cloak–attachment>
44ρ¹	24♐43	+4;30	3.92		The middle one of these
46υ	24♐59	+6;21	4.52		The northern one of the three
54e¹	29♐25	+5;17	5.3		The dim one east of these three
55e²	29♐52	+5;24	5.05		
61g	3♉40	+5;16	5.00		The northern one of the two in the southern <cloak–attachment>
56f	0♉14	+1;36	4.88		The southern one of these
47χ¹	24♐32	−2;15	5.02		The star in the right shoulder
49χ³	24♐43	−1;42	5.44		
51h¹	26♐57	−2;50	5.64		The star in the right elbow
52h²	27♐03	−3;00	4.59		
42ψ	22♐15	−2;40	4.85		Of the three in the back, the star <just above the place between the shoulders>
40τ	20♐05	−4;58	3.31		The middle one of these <just above the shoulder–blade>
38ζ	18♐53	−6;55	2.60	Ascella	The remaining one below the arm-pit
54β¹	20♐59	−21;53	3.91	Arkab or Urkab	The star in the left fore–ankle

APPENDIX I 177

CAT.	long.	LAT.	MAG.	NAME	DESCRIPTION
62β²	20♐59	−22;15	4.28		
68α	21♐49	−18;11	3.96	Rukbat or Al Rami	The star in the knee of the same <leg>
17η	8♐55	−13;12	3.07		The star in the right fore–ankle
330θ¹	0♑04	−14;09	4.36		The star in the left thigh
333θ²	0♑10	−13;36	5.30		
297ι	27♐45	−20;22	4.11		The star in the right <hind lower leg>
58ω	0♑59	−5;06	4.69	Terebellum	The western star of the northern side of the four at the beginning of the tail
60A	1♑46	−5;11	4.83		The eastern star of the northern side
59b	1♑09	−6;05	4.53		The western star of the southern side
61c	2♑16	−6;51	4.50		The eastern star of the southern side

CAPRICORNUS
(Constellation of the Goat)

CAT.	LONG.	LAT.	MAG.	NAME	DESCRIPTION
5α¹	9♑01	+7;13	4.30	Algedi or Algiedi	The northern star of the three in the eastern horn
6α²	9♑05	+7;10	3.57		
8ν	9♑41	+6;48	4.75	Al Shat	The middle star of these
9β	9♑17	+4;50	3.08	Dabih	The southern one of the three
1ξ¹	7♑42	+7;38	6.33		The star at the tip of the western horn
2ξ²	7♑37	+7;19	5.85		
12o	10♑26	+0;35	5.93		The southern star of the three in the muzzle
10π	9♑57	+1;07	5.10		The western one of the remaining two
11ρ	10♑25	+1;25	4.76		The eastern one of these
7σ	7♑55	+0;41	5.28		The western one of the three under the right eye
13τ¹	13♑32	+3;33	5.23		The northern one of the two in the neck
14τ²	13♑32	+3;33	5.65		
15υ	12♑55	+0;26	5.16		The southern one of these
16ψ	12♑23	−6;54	4.13		The star under the right knee
18ω	13♑11	−8;45	4.10		The star in the left bent knee
24A	17♑04	−7;55	4.49		The star in the left shoulder
34ζ	22♑09	−6;47	3.74		The western star of the two close together under the belly
36b	22♑44	−6;22	4.50		The eastern one of these

CAT.	LONG.	LAT.	MAG.	NAME	DESCRIPTION
28φ	20♉15	−4;20	5.18		The eastern one of the three in the middle of the body
25χ	18♉29	−4;23	5.29		The southern one of the two remaining western stars
22η	17♉59	−2;49	4.82		The northern one of these
23θ	19♉02	−0;24	4.06		The western one of the two in the back
32ι	22♉54	−1;10	4.28		The eastern one of these
39ε	25♉25	−4;48	4.60		The western one of the two in the southern part of the <fish–tail>
43κ	26♉47	−4;39	4.72		The eastern one of these
40γ	26♉55	−2;22	3.67	Nashira	The western star of the two near the tail
49δ	28♉35	−2;33	2.85	Deneb Algedi	The eastern one of these
42d	28♉13	−0;14	5.15		The western one of the four in the northern part of the tail
51μ	0♒54	−0;28	5.07		The southern one of the remaining three
48λ	0♒14	+2;05	5.56		The middle one of these
46c¹	0♒39	+4;22	5.09		The northern one of these <on the end of the tail–fin>

AQUARIUS
(Constellation of the Waterman or Water Bearer)

CAT.	LONG.	LAT.	MAG.	NAME	DESCRIPTION
25d	3♒17	+15;32	5.10		The star in the head of the Water Bearer
34α	8♒38	+10;47	2.94	Sadalmelik	The brighter of the two in the right shoulder
31o	7♒22	+9;18	4.78	Al Sa'd al Mulk	The dimmer one beneath this
22β	28♉39	+8;47	2.89	Sadalsuud	The star in the left shoulder
23ξ	29♉19	+6;08	4.68		The star under it in the back, as if under the armpit
13ν	21♉36	+4;58	4.50		The eastern one of the three in the left <arm, on the coat>
6μ	18♉18	+8;26	4.72		The middle one of these
2ε	16♉58	+8;16	3.77	Al Bali	The western one of the three
48γ	11♒56	+8;23	3.84	Sadachbia	The star in the right forearm
52π	13♒53	+10;35	4.56	Seat	The northern one of the three <on the right hand>
55ζ¹	14♒05	+9;01	3.64		The western one of the two remaining <southern> ones
55ζ²	14♒06	+9;00	4.42		
62η	15♒37	+8;14	4.02		The eastern one of these

APPENDIX I

CAT.	LONG.	LAT.	MAG.	NAME	DESCRIPTION
43θ	8♒27	+2;51	4.17	Ancha	The western star of the two close together in the <hollow of the right [hip]>
46ρ	9♒16	+2;30	5.35		The eastern one of these
57σ	10♒37	−1;08	4.82		The star in the right buttock
33ι	3♒55	−1;58	4.26		The southern star of the two in the left buttock
38e	5♒43	−0;08	5.42		The northern one of these
76δ	14♒05	−8;07	3.27	Skat or Scheat	The southern star of the two in the right <lower leg>
71τ²	13♒48	−5;35	4.01		The northern one of these below the bend of the knee
53f	7♒18	−6;20	5.55		The star in the left <thigh>
68g²	11♒00	−11;03	5.24		The southern one of the two in the left <lower leg>
66g¹	10♒26	−9;52	4.67		The northern one of these under the knee
63κ	14♒41	+4;08	5.04	Situla	The western one of these in the flow of water from the hand
73λ	16♒49	−0;17	3.74		The star near this last southwards
83h	19♒34	−1;35	5.42		The star near this last after the curve
90φ	22♒18	−1;05	4.22		The star still east of this
92χ	22♒18	−2;47	4.97		The star in the curve south of this last
91ψ¹	21♒20	−3;52	4.23		The northern one of the two south of this
93ψ²	21♒56	−4;13	4.39		The southern one of the two
95ψ³	22♒00	−4;42	4.98		
94*	20♒20	−8;15	5.20		The lone star distant from these towards the south
102ω¹	24♒49	−11;00	4.97		The western star of the two close together after this one
105ω²	25♒19	−11;35	4.48		The eastern one of these
103A¹	23♒39	−14;41	5.34		The northern one of the three in the nearby stream
104A²	23♒47	−14;27	4.81		
106i¹	24♒07	−15;07	5.23		The middle one of the three
108i³	25♒27	−16;24	5.18		The eastern one of these
98b¹	18♒40	−14;47	3.95		The northern one of the next three, in like manner
99b²	19♒05	−15;32	4.38		The middle one of these
101b³	20♒34	−16;27	4.69		The southern one of the three
86c¹	13♒28	−16;28	4.48		The western one of the three in the remaining stream

CAT.	LONG.	LAT.	MAG.	NAME	DESCRIPTION
89c³	14≈44	−15;37	4.71		The southern one of the remaining two
88c²	15≈10	−14;22	3.67		The northern one of these
79**	8≈48	−21;02	1.16	Fomalhaut	The last star in the water and in the southern Fish's mouth

Unfigured Stars around the Waterman or Water Bearer

CAT.	LONG.	LAT.	MAG.	NAME	DESCRIPTION
2⁺	28≈55	−16;12	4.55		The western star of the three stars east of the water's curve
6⁺	1)(25	−15;25	4.89		The northern one of the remaining two
7⁺	0)(41	−18;47	4.43		The southern one of these

*94 Aquarii is the choice of Peters and Knobel. Toomer (1984), p378 chooses BSC 8958
** = α Piscis Austrini;   ⁺ Ceti

PISCES
(Constellation of the Fishes)

CAT.	LONG.	LAT.	MAG.	NAME	DESCRIPTION
4β	23≈52	+9;07	4.53	Fum al Samakah	The star in the mouth of the western Fish
6γ	26≈19	+7;28	3.69		The southern one of the two in the top of his head
7b	28≈17	+8;57	5.04		The northern one of these
10θ	0)(33	+9;03	4.27		The western one of the two in the back
17ι	2)(49	+7;29	4.12		The eastern one of these
8κ	28≈08	+4;32	4.93		The western one of the two in the belly
18λ	1)(57	+3;29	4.50		The eastern one of these
28ω	7)(48	+6;27	4.00		The star in the tail of the same Fish
41d	13)(15	+5;26	5.36		The first star from the tail, in the cord
51	15)(25	+3;08	5.68		The eastern one of these
63δ	19)(22	+2;09	4.42		The western star of the three bright ones following
71ε	22)(48	+0;59	4.27		The middle one of these
86ζ	25)(03	−0;15	4.85	Revati	The eastern one of the three
80e²	23)(20	−1;32	5.52		The northern one of the two little ones in the curve under these
89f	24)(34	−4;21	5.16		The southern one of these
98μ	28)(13	−3;05	4.84		The western one of the three after the curve

APPENDIX I   181

CAT.	LONG.	LAT.	MAG.	NAME	DESCRIPTION
106ν	0♈44	−4;49	4.44		The middle one of these
111ξ	2♈43	−8;03	4.61		The eastern one of the three
113α	4♈34	−9;11	4.33	Al Rescha or Al Rischa	The star in the knot of the two cords
110ο	2♈56	−1;46	4.26	Torcular	The star west of the knot, in the northern cord
102π	2♈11	+1;44	5.55		The southern one of the next three after it
99η	2♈04	+5;16	3.61	Alpherg	The middle one of these
93ρ	2♈22	+9;15	5.32		The northern one of the three, and at the tail's end
94	2♈28	+9;20	5.49		
82g	4♈08	+21;53	5.15		The northern one of the two in the mouth of the eastern Fish
83τ	3♈37	+20;38	4.51		The southern one of these
68h	0♈14	+20;51	5.41		The eastern of the three little ones in the head
67k	29♓03	+19;23	6.06		The middle one of these
65i	27♓56	+20;26	6.30		The western one of the three
74ψ¹	28♓42	+13;17	5.34		The western star of the three in the fin <spine> after the star in Andromeda's elbow
79ψ²	28♓55	+11;12	5.57		The middle one of these
81ψ³	28♓42	+13;16	5.56		The eastern one of the three
90υ	4♈06	+17;21	4.75		The northern one of the two in the belly
85φ	1♈46	+15;24	4.65		The southern one of these
84χ	29♓49	+12;20	4.66		The star in the eastern fin <spine> near the tail

Unfigured Stars around the Fishes

27	3♓31	−3;10	4.86		The western star of the two northern ones of the square under the western Fish
29	4♓26	−2;57	5.11		The eastern one of these
30	3♓13	−5;43	4.39		The western star of the southern side
33	4♓11	−5;43	4.61		The eastern one of the southern side

Identification of ψ³ and χ (Peters and Knobel) is reversed by Toomer (1984), pp 380–381

CETUS
(Constellation of the Sea-Monster)

CAT.	LONG.	LAT.	MAG.	NAME	DESCRIPTION
91λ	20 ♈ 18	−7;58	4.69		The star at the nostril tip
92α	19 ♈ 32	−12:44	2.52	Menkar or Menkab	The eastern star of the three in the muzzle at the tip of the jaw
86γ	14 ♈ 44	−12;07	3.46	Al Kaff al Jidhmah	The middle one of these in the middle of the mouth
82δ	12 ♈ 45	−14;38	4.07		The western one of the three, in the cheek
65ξ¹	9 ♈ 17	−4;25	4.36		The star west of the brow and eye, in the flowing hair
72ρ	4 ♈ 51	−25;23	4.88		The northern star of the western side of the square in the chest
76σ	5 ♈ 14	−28;45	4.74		The southern star of the western side
83ε	8 ♈ 19	−26;15	4.82		The northern star of the eastern side
89π	8 ♈ 52	−28;25	4.23		The southern star of the eastern side
52τ	24 ♓ 08	−24;50	3.49		The middle one of the three in the body
59υ	24 ♓ 25	−31;07	3.99		The southern one of these
55ζ	27 ♓ 04	−20;27	3.72	Baten Kaitos	The northern one of the three
45θ	21 ♓ 23	−15;59	3.60		The eastern one of the two near the tail
31η	16 ♓ 47	−16;12	3.44	Dheneb	The western one of these
19φ²	12 ♓ 29	−14;56	5.18		The northern star of the eastern side of the square near the tail
22φ³	13 ♓ 33	−15;56	5.32		The southern one of the eastern side
17φ¹	11 ♓ 02	−14;12	4.75	Al Nitham	The northern star of the western side
23φ⁴	14 ♓ 08	−16;19	5.61		The southern one of the western side
8ι	6 ♓ 06	−10;03	3.55	Schemali	The northern star of the two at the tail's tip <tail–fin>
16β	7 ♓ 36	−20;43	2.03	Deneb Kaitos	The star at the <end of the southern tail–fin>

The two southern stars of the square are identified as φ³ and φ⁴ by Peters and Knobel and also by Toomer (1984), p 382, where φ³ = BSC 227 and φ⁴ = BSC 190

ORION
(Constellation of the Hunter or Warrior)

CAT.	LONG.	LAT.	MAG.	NAME	DESCRIPTION
39λ	28 ♉ 56	−13;39	3.39	Meissa	The nebula in Orion's head
58α	3 ♊ 59	−16;19	0.87	Betelgeuze Betelgeuse	The bright red star in the right shoulder

APPENDIX I   183

CAT.	LONG.	LAT.	MAG.	NAME	DESCRIPTION
24γ	26 ♉ 11	−17;05	1.63	Bellatrix	The star in the left shoulder
32A	27 ♉ 37	−17;33	4.20		The eastern star below this
61μ	5 ♊ 51	−14;03	4.12		The star in the right elbow
74k	9 ♊ 18	−11;30	5.04		The star in the right forearm
70ξ	8 ♊ 10	−9;28	4.47		The eastern double star of the southern side of the square in the right hand
67ν	7 ♊ 05	−8;55	4.41		The western star of the southern side
72f²	8 ♊ 58	−7;31	5.33		The eastern star of the northern side
69f¹	8 ♊ 10	−7;32	4.95		The western star of the northern side
54χ¹	4 ♊ 01	−3;23	4.40		The western star of the two in the club <staff>
62χ²	6 ♊ 10	−3;34	4.63		The eastern one of these
47ω	29 ♉ 44	−19;30	4.54		The eastern one of the 4 stars in a straight line in the back
38n²	28 ♉ 26	−19;47	5.35		The star west of this one
33n¹	27 ♉ 35	−20;14	5.45		The star still west of this last
30ψ²	26 ♉ 24	−20;21	4.58		The last and western one of the four
15y²	23 ♉ 01	−7;34	4.81		The northern one of those in <the pelt on the left arm>
11y¹	21 ♉ 46	−7;38	4.68		The second one from the northernmost
9o²	19 ♉ 37	−9;19	4.06		The third from the northernmost
7π¹	18 ♉ 48	−12;33	4.64		The fourth from the northernmost
2π²	17 ♉ 36	−13;44	4.35		The fifth from the northernmost
1π³	16 ♉ 52	−15;37	3.18		The sixth from the northernmost
3π⁴	17 ♉ 19	−17;02	3.68		The seventh from the northernmost
8π⁵	17 ♉ 42	−20;16	3.70		The eighth from the northernmost
10π⁶	18 ♉ 45	−21;07	4.47		The last and southernmost of those in the skin <pelt>
34δ	27 ♉ 35	−23;50	2.25	Mintaka	The western star of the three in the belt
46ε	28 ♉ 41	−24;47	1.69	Alnilam	The middle one of these
50ζ	29 ♉ 54	−25;34	1.76	Alnitak	The eastern one of the three
28η	25 ♉ 22	−25;48	3.35	Algiebba	The star in the handle of the dagger
42	28 ♉ 16	−28;24	4.58		The northern one of the three bunched at the tip of the dagger
41θ¹	28 ♉ 12	−28;57	5.13		The middle one of these
43θ²	28 ♉ 14	−28;59	5.07		
44ι	28 ♉ 13	−29;29	2.77	Nair al Saif	The southern one of the three
49d	29 ♉ 09	−30;51	4.79		The eastern star of the two under the tip of the <dagger>
36υ	27 ♉ 08	−30;49	4.61	Thabit	The western one of these

CAT.	LONG.	LAT.	MAG.	NAME	DESCRIPTION
19β	22♉02	−31;24	0.10	Rigel	The bright star in the left foot common with the Water
20τ	23♉04	−30;07	3.59		The northern one of those <in the lower leg, over the ankle–joint>
29e	24♉48	−31;13	4.12		The star outside under the left heel
53κ	1♊38	−33;21	2.06	Saiph	The star under the right and eastern knee

ERIDANUS
(Constellation of the River)

CAT.	LONG.	LAT.	MAG.	NAME	DESCRIPTION
69λ	20♉24	−31;49	4.25		The star after the star in the foot of Orion at the beginning of the River
67β	20♉32	−28;11	2.78	Cursa	The star north of this one in the bend near Orion's shin
65ψ	18♉24	−30;02	4.80		The eastern one of the two after this one
61ω	16♉15	−28;03	4.38		The western one of these
57μ	14♉31	−25;38	4.01		Again the eastern one of the next two
48ν	12♉00	−25;23	3.93		The western one of these
42ξ	8♉31	−25;16	5.17		The eastern one of the three after this last
40o²	6♉21	−30;52	4.42	Keid	The middle one of these
38o¹	4♉36	−27;39	4.04	Beid	The western one of the three
34γ	28♈56	−33;29	2.95	Zaurak	The eastern star of the four in the next interval
26π	26♈05	−31;18	4.42		The star west of this one
23δ	26♈11	−28;29	3.52	Rana	The star west again of this last
18ε	23♈55	−28;04	3.72		The western one of the four
13ζ	18♈59	−26;05	4.79		Likewise the eastern one of the four in the next interval
9ρ²	15♈54	−24;03	5.30		The star west of this last
10ρ³	16♈19	−24;05	5.25		
3η	13♈48	−24;50	3.89	Azha	The star again west of this
1τ¹	7♈00	−32;49	4.46		The first star in the bend of the River and touching the chest of the Sea–Monster
2τ²	7♈44	−35;41	4.76	Argentenar	The star east of this
11τ³	9♈40	−39;07	4.08		The western one of the next three
16τ⁴	15♈09	−38;39	3.65		The middle one of these
19τ⁵	19♈14	−39;38	4.26		The eastern one of the three
27τ⁶	22♈25	−42;25	4.22		The northern one of the western side of the four in a trapezium

APPENDIX I

CAT.	LONG.	LAT.	MAG.	NAME	DESCRIPTION
28τ⁷	22♈24	−42;43	5.23		The southern one of the western side
33τ⁸	23♈55	−43;52	4.63		The western one of the eastern side
36τ⁹	26♈02	−43;41	4.64		The eastern one of this side and last of the four
50υ¹	4♉35	−51;21	4.50		The northern one of the two distant stars to the east
52υ²	4♉59	−52;04	3.81		The southern one of these
43υ³	29♈27	−54;44	3.96		The eastern one of the next two after the turn
41υ⁴	27♈27	−54;11	3.54		The western one of these
202*	18♈57	−53;26	5.09		The eastern one of the three in the next interval
189*	16♈43	−54;32	4.16		The middle one of these
149*	13♈53	−55;04	4.58		The western one of the three
θ¹ ‡	28♓12	−53;51	2.90	Acamar	The last and bright star of the River

* III.   ‡ This star (θ¹) was known to Arabian astronomers as the star at the end of the River and was originally called by them Achernar, which means '*the End of the River*'. Then the name Achernar was appropriately given to α Eridani, which is by far the brightest star (magnitude m = 0.45) in this constellation and is truly at the end of the river (sidereal longitude ♒20°). The star θ¹ was subsequently called Acamar. It is curious that Ptolemy does not mention Achernar in his catalogue, since—despite the extreme southerly latitude of Achernar (declination −59°18′)—he could have seen it from the latitude of Alexandria. Achernar's ecliptic coordinates in the Babylonian sidereal zodiac for the epoch −100 were: 20♒02 and −59;18

LEPUS
(Constellation of the Hare)

CAT.	LONG.	LAT.	MAG.	NAME	DESCRIPTION
3ι	20♉56	−35;00	4.45		The northern star of the western side of the square <just> over the ears
4κ	21♉06	−36;05	4.36		The southern one of the western side
7ν	23♉12	−35;36	5.29		The northern one of the eastern side
6λ	22♉59	−36;28	4.29		The southern one of the eastern side
5μ	20♉33	−39;20	3.19		The star in the <cheek>
2ε	17♉11	−45;16	3.17		The star in the left forefoot
11α	26♉35	−41;20	2.58	Arneb	The star in the middle of the body
9β	24♉52	−44;14	2.83	Nihal	The star under the belly
15δ	2♊12	−44;57	3.77		The northern one of the two in the hind <legs>
13γ	0♊17	−46;19	3.59		The southern one of these
14ζ	1♊13	−38;30	3.54		The star in the loin
16η	4♊10	−37;48	3.71		The star at the tail's tip

CANIS MAJOR
(Constellation of the Dog)

CAT.	LONG.	LAT.	MAG.	NAME	DESCRIPTION
9α	19♊53	−40;33	−1.44	Sirius	The brightest and red star in the <mouth>, called the Dog
14θ	21♊34	−34;59	4.07		The star in the ears
18μ	22♊21	−36;55	4.97		The star in the head
23γ	24♊55	−38;16	4.10	Muliphen	The northern one of the two in the neck
20ι	22♊50	−39;54	4.38		The southern one of these
15†	22♊34	−43;08	4.82		The star in the chest
8ν³	17♊18	−41;34	4.42		The northern one of the two in the right knee
7ν²	17♊00	−42;38	3.95		The southern one of these
2β	12♊28	−41;32	1.97	Murzim	The star at the end of <the front leg>
4ξ¹	15♊57	−46;50	4.33		The western star of the two in the left knee
5ξ²	16♊57	−46;19	4.43		The eastern one of these
24ο²	26♊21	−46;23	3.03		The eastern star of the two in the left shoulder
16ο¹	23♊30	−47;02	3.88		The western one of these
25δ	28♊46	−48;42	1.83	Wezen	The star at the beginning of the left thigh
21ε	26♊07	−51;37	1.50	Adhara	The star under the belly between the thighs
13κ	23♊56	−55;25	3.90		The star in the joint of the right <leg>
1ζ	12♊40	−53;39	3.02	Furud	The star at the tip of the right <leg>
31η	4♋57	−50;51	2.44	Aludra	The star in the tail

The Unfigured Stars about the Dog

CAT.	LONG.	LAT.	MAG.	NAME	DESCRIPTION
22δ*	24♉49	−22;59	4.15		The star north of the Dog's head
9θ⁺	8♊19	−60;57	5.00		The southernmost one of the four in a straight line under the hind <legs>
65κ⁺	11♊46	−58;43	4.36		The star north of this
95δ⁺	13♊45	−57;01	3.84		The star still north of this
136λ	15♊54	−56;01	4.47		The last and northern star of the four
238μ⁺	29♉56	−55;58	5.17		The western star of the three in a straight line <west> of these four
276λ⁺	2♊35	−57;29	4.87		The middle one of these
297γ⁺	4♊16	−59;00	4.32		The eastern one of the three

CAT.	LONG.	LAT.	MAG.	NAME	DESCRIPTION
267β⁺	1♊35	−59;13	3.11	Wezn or Wazn	The eastern star of the two bright ones under these
196α⁺	27♉20	−57;40	2.65	Phaet	The western star of these
140ε⁺	23♉49	−58;55	3.86		The last and southern star of the foregoing

†This star is identified as π Canis Majoris by Toomer (1984), p 387
*Monocerotis;   +Columbae

CANIS MINOR
(Constellation of Procyon — the Little Dog)

CAT.	LONG.	LAT.	MAG.	NAME	DESCRIPTION
3β	27♊29	−13;43	2.89	Gomeisa	The star in the neck
10α	1♋22	−15;37	0.40	Procyon	The bright star in the hinder parts called Procyon

ARGO NAVIS
(Constellation of the Ship Argo)

CAT.	LONG.	LAT.	MAG.	NAME	DESCRIPTION
11e*	13♋03	−42;47	4.19		The western star of the two in <the stern−ornament>
15ρ*	16♋51	−43;26	2.80		The eastern one of these
7ξ*	11♋27	−45;10	3.34		The northern one of the two close together above the Shield in the stern
2200*	11♋29	−46;16	4.47		The southern one of these
173m*	8♋11	−46;17	4.68		The star west of these
BSC 2948⁺	8♋52	−47;38	3.82		The bright star in the middle of the Shield
163p*	8♋24	−49;21	4.64		The western star of the three below the Shield
3*	11♋19	−49;26	3.96		The eastern one of these
1*	11♋01	−48;54	4.58		The middle one of the three
BSC 3113	16♋23	−49;53	4.78		The star in the goose neck of the stern
99′	5♋35	−53;27	5.41		The northern one of the two in the stern keel
108′	6♋08	−53;15	5.34		
68π*	5♋47	−58;46	2.70		The southern one of these
172f*	11♋58	−55;35	4.53		The northern one of those in the deck of the poop
186d″	14♋36	−58;38	4.82		The western one of the next three
214c*	16♋25	−57;56	3.60		The middle one of these
254b*	19♋34	−58;16	4.49		The eastern one of the three

188   HISTORY OF THE ZODIAC

CAT.	LONG.	LAT.	MAG.	NAME	DESCRIPTION
306ζ*	24♋09	−58;31	2.24	Naos	The bright one east of these, on the deck
253a*	20♋40	−59;54	3.71		The western one of the two dim stars under the bright one
BSC 3162	24♋44	−62;13	5.36		The eastern one of these
21h¹*	26♋25	−57;35	4.44		The <western> star of the two above the bright one
35h²*	27♋52	−58;05	4.43		The <eastern> one of these
BSC 3439	8♌52	−54;58	5.48		The northern star of the three in the shields near the mastholder
168d‡	9♌20	−57;28	4.02		The middle one of these
139e‡	7♌36	−58;23	4.14		The southern one of the three
176a‡	13♌11	−60;14	3.90		The northern one of the two close together under these
155b‡	12♌15	−61;15	3.80		The southern one of these
145β†	2♌20	−51;19	3.96		The southern one of the two in the middle of the mast
162α†	2♌01	−49;04	3.67		The northern one of these
193γ†	0♌59	−43;23	4.01		The western star of the two near the tip of the mast
220δ†	2♌18	−43;05	4.87		The eastern one of these
1λ‡	16♌49	−55;57	2.20		The star below the third and eastern shield
116ψ‡	20♌23	−51;09	3.60		The star on the <cut–off> of the deck
135σ*	14♋15	−63;53	3.21		The star between the oars in the keel
235P*	24♋23	−65;45	5.21		The dim star east of this
γ²‡	3♌05	−64;37	1.79	Suhail	The bright star east of this below the deck
χΔ	6♌39	−70;27	3.45		The bright star south of this in the lower keel
o‡	20♌43	−66;29	5.17		The western star of the three east of this
δ‡	24♌54	−67;17	1.95		The middle one of these
fΔ	29♌14	−68;30	4.47		The eastern one of the three
κ‡	4♍42	−63;43	2.49	Markab	Of the two east of these, the western one near the <cut–off>
N‡	10♍02	−64;12	3.14		The eastern one of these
315η∩	4♊49	−66;32	3.95		The western one of the two in the northern and western oar
205v*	22♊34	−66;21	3.17		The eastern one of these
αΔ	20♊27	−76;05	−0.73	Canopus	Of the two in the remaining oar the western star called Canopus
τ*	3♋23	−73;09	2.93		The last and eastern one of these

* Puppis;   ′VII Puppis;   ″VII Puppis;   ‡ Velorum;   † Pyxidis;
∩ Columbae;   Δ Carinae;   + BSC 2948 + BSC 2949, cf. Toomer (1984), p 389

HYDRA
(Constellation of the Water Snake)

CAT.	LONG.	LAT.	MAG.	NAME	DESCRIPTION
5σ	16♋30	−14;48	4.44	Minhar al Shuja	Of the two western stars in the head, the southern one in the nostrils
4δ	15♋37	−12;35	4.14		The northern one of these, above the eye
11ε	17♋43	−11;15	3.38		Of the two stars east of these, the northern one <which is about on the skull>
7η	17♋36	−14;27	4.29		The southern one of these, in the <gaping jaws>
16ζ	19♋55	−11;10	3.10		The star east of them all <about on the cheek>
18ω	22♋41	−11;13	4.97		The western one of the two at the beginning of the neck
22θ	25♋27	−13;05	3.88		The eastern one of these
32τ²	1♌02	−15;08	4.56	Ukdah	The middle one of the next three in the curve of the neck
35ι	2♌56	−14;28	3.90		The eastern one of the three
31τ¹	0♌49	−16;54	4.59		The southernmost of these
BSC 3750	1♌52	−20;05	5.37		The dim northern star of the two close together in the south
30α	2♌37	−22;31	1.97	Alphard	The bright one of these two
38κ	8♌04	−26;44	5.04		The western star of the three east of the curve
39υ¹	11♌03	−26;13	4.10		The middle one of these
40υ²	13♌42	−23;16	4.58		The eastern one of the three
42μ	20♌30	−24;45	3.81		The western one of the next three almost in a straight line
φ³	23♌27	−23;31	4.90		The middle one of these
ν	25♌37	−21;46	3.11		The eastern one of the three
11β*	3♍57	−25;42	4.47		The northern one of the two after the base of the Bowl
χ¹	4♍49	−30;14	4.91		The southern one of these
ξ	13♍31	−31;32	3.54		The western one of the three in the triangle after these
o	16♍37	−33;23	4.70		The middle and southern one of these
β	18♍53	−31;23	4.28		The eastern one of the three
46γ	2♎18	−13;39	2.99		The star behind the Raven near the tail
49π	13♎55	−12;58	3.26		The star at the tip of the tail
BSC 3314	15♋13	−22;40	3.89		<The star to the south of the head>
15α‡	9♌25	−11;14	4.48		The star far east of those in the neck

*Crateris   ‡ Sextantis

CRATER
(Constellation of the Bowl)

CAT.	LONG.	LAT.	MAG.	NAME	DESCRIPTION
7α	29♌16	−22;34	4.08	Alkes	The star in the base of the Bowl common with the Water–Snake
15γ	4♍37	−19;39	4.08		The southern one of the two in the middle of the Bowl
12δ	2♍02	−17;27	3.55		The northern one of these
27ζ	9♍24	−18;18	4.72		<The star on the southern rim of the mouth>
14ε	1♍34	−13;28	4.82		The star on the northern edge
30η	11♍27	−16;03	5.17		The star on the southern handle
21θ	3♍55	−11;18	4.69		The star on the northern handle

CORVUS
(Constellation of the Raven)

CAT.	LONG.	LAT.	MAG.	NAME	DESCRIPTION
1α	17♍33	−21;44	4.02	Al Chiba	The star in the beak and common with the Water–snake
2ε	17♍02	−19;36	3.00	Minkar	The star in the neck near the head
5ζ	19♍11	−18;13	5.21		The star in the breast
4γ	16♍07	−14;24	2.58	Gienah	The star in the western and right wing
7δ	18♍54	−12;09	2.94	Algorab	The western star of the two in the eastern wing
8η	19♍21	−11;33	4.30		The eastern one of these
9β	22♍42	−17;59	2.64	Kraz	The star at the end of the <leg>, common with the Water–snake

CENTAURUS
(Constellation of the Centaur)

CAT.	LONG.	LAT.	MAG.	NAME	DESCRIPTION
2g	13♎23	−21;28	4.19		Southernmost of the four in the head
4h	13♎07	−18;49	4.74		The northernmost of these
1i	12♎16	−20;26	4.21		The western star of the remaining two middle ones
3k	13♎16	−19;55	4.56		The eastern and last one of these four
53ι*	8♎40	−25;51	2.75		The star on the left and western shoulder
5θ	18♎00	−22;05	2.05	Menkent	The star on the right shoulder
99d*	11♎51	−27;27	3.88		The star on the left shoulder blade
40ψ**	21♎02	−22;18	4.04		The northern star of the two western ones of the four in the <thyrsus> wand
55a**	22♎08	−23;39	4.39		The southern one of these

APPENDIX I 191

CAT.	LONG.	LAT.	MAG.	NAME	description
150c¹**	24♎45	−18;16	4.04		Of the remaining two, the star at the tip of the <thyrsus> wand
141b**	25♎15	−20;47	3.99		The remaining one south of this
197v*	16♎31	−28;05	3.40		The western one of the three in the right side
198μ*	16♎54	−28;48	3.19		The middle one of these
246φ*	18♎24	−27;49	3.81		The eastern one of the three
288χ*	19♎30	−26;24	4.36		The star in the right arm
109η**	25♎36	−25;19	2.35		The star in the right forearm
216κ**	0♏07	−23;49	3.13		<The star in the right hand>
231ζ*	20♎21	−32;46	2.54		The bright star at the beginning of the human body
267υ²*	20♎39	−30;47	4.32		The eastern star of the two dim stars north of this one
249υ¹*	19♎43	−30;16	3.86		The western one of these
Cum.ω	15♎05	−36;18	6.36		The star at the beginning of the back
f	11♎59	−37;33	4.70		The star west of this last on the horse's back
γ	7♎50	−39;59	2.20	Muhlifain	The eastern star of the three <on the rump>
τ	6♎51	−39;57	3.87		The middle one of these
σ	6♎12	−42;15	3.90		The western one of the three
δ	2♎59	−44;23	2.57		The western star of the two close together in the right thigh
ρ	4♎54	−45;27	3.96		The eastern one of these
M	20♎56	−37;08	4.64		The star in the chest under the horse's armpit
ε	20♎58	−39;23	2.30		The western star of the two under the belly
Q	21♎57	−40;14	4.99		The eastern one of these
γ⁺	12♎21	−47;47	1.62	Gacrux	<The star on the knee–bend of the right [hind] leg>
β⁺	17♎09	−48;27	1.25	Mimosa	<The star in the hock of the same leg>
δ⁺	11♎12	−50;15	2.79		<The star under the knee–bend of the left [hind] leg>
α⁺	17♎24	−52;41	0.77	Acrux	<The star on the frog of the hoof on the same leg>
α¹	5♏54	−41;34	−0.01	Rigil or Toliman	<The star at the end of the right front leg>
α²	5♏54	−41;34	1.35		
β	29♎13	−43;55	0.61	Agena or Hadar	<The star on the knee of the left [front] leg>
μ¹⁺	16♎05	−45;53	4.03		<The star outside, under the right hind leg>
μ²⁺	16♎05	−45;53	5.08		

*XIII;   ** XIV;   ⁺ Crucis

LUPUS
(Constellation of the Wild Beast or Wolf)

CAT.	LONG.	LAT.	MAG.	NAME	DESCRIPTION
211β*	0♏22	−24;50	2.67		The star at the end of the hind foot near the Centaur's hand
α	28♎51	−29;49	2.3	Men	The star in the hand of the same foot
31δ**	3♏58	−21;12	3.21		The western star of the two on the shoulder blade
98γ**	6♏48	−21;01	2.80		The eastern one of these
35ε**	5♏26	−25;01	3.37		The star in the middle of the Wild Beast's body
λ	3♏02	−26;18	4.05		The star in the belly under the flank
242π^a**	2♏58	−28;11	3.89		The star in the thigh
μ	5♏42	−28;17	4.27		The northern one of the two near the beginning of the thigh
κ^1	4♏50	−29;26	3.88		The southern one of these
ζ	6♏09	−32;37	3.40		The star at the end of the loins
ρ	28♎59	−31;57	4.03		The southern star of the three at the tip of the tail
ι	24♎08	−29;59	3.55		The middle one of the three
66τ^2*	25♎06	−28;55	4.35		The northern one of these
217η**	11♏03	−17;12	3.42		The southern star of the two in the neck
248θ**	12♏02	−15;23	4.21		The northern one of these
174χ	8♏07	−12;57	3.95		The western one of the two in the muzzle
204ξ^1**	9♏25	−13;01	5.13		The eastern one of these
1i	29♎58	−12;47	4.90		The southern one of the two in the forefoot
2f	0♏17	−11;17	4.33		The northern one of these

*XIV;   **XV;   instead of χ Lupi and ξ^1 Lupi, Toomer (following Manitius) identifies these two stars as ψ^1+ψ^2 Lupi and χ Lupi (1984), pp 396–397

ARA
(Constellation of the Censer)

CAT.	LONG.	LAT.	MAG.	NAME	DESCRIPTION
σ	0♐43	−22;54	4.57		The northern star of the two in the base
θ	6♐26	−26;23	3.67		The southern one of these
α	0♐12	−26;19	2.87		\<The star in the middle of the little altar\>
ε^1	24♏51	−29;59	4.05		The northern one of the three in the brazier
γ	29♏33	−32;50	3.32		The southern one of the remaining two contiguous ones
β	29♏28	−32;00	2.84		The northern one of these
ζ	25♏06	−32;50	3.12		The star at the end of the burning flame

APPENDIX I 193

CORONA AUSTRALIS
(Constellation of the Southern Crown)

CAT.	LONG.	LAT.	MAG.	NAME	DESCRIPTION
73δ^{1+}	11♐12	−22;23	4.93		The outside western star of the southern edge
76δ^{2+}	11♐16	−22;13	5.08		
166η^{1}*	14♐37	−20;22	5.47		The eastern star of those on the Crown
169η^{2}*	14♐48	−20;09	5.60		
BSC 7122	16♐10	−19;33	5.34		The star east of this one
25oζ*	17♐32	−19;05	4.74		The star again east of this one
291δ*	18♐46	−17;36	4.58		The star after this one in front of the Archer's <knee>
305β*	19♐16	−16;30	4.10		The star after this one and north of the bright one in the knee
300α*	19♐19	−15;06	4.10		The star north of this one
28oγ*	18♐47	−14;17	4.23		The star still north of this one
23oε*	17♐19	−14;04	4.87		The eastern one of the two western ones following this last on the northern edge
BSC 7129	16♐49	−14;13	5.36		The western one of these two dim ones
142λ*	14♐06	−14;59	5.10		The star rather west of this one
BSC 6942	11♐42	−16;12	5.16		The star still west of this
85θ*	11♐45	−18;48	4.63		The last one south of this last

+XVIII Telescopii;   *XVIII

PISCIS AUSTRINUS
(Constellation of the Southern Fish)

CAT.	LONG.	LAT.	MAG.	NAME	DESCRIPTION
24α	8♒48	−21;02	1.16	Fomalhaut	The star in the mouth, the same as that at the beginning of the Water
17β	2♒18	−21;13	4.28		The western star of three at the southern edge of the head
22γ	6♒28	−23;32	4.46		The middle one of these
23δ	7♒20	−23;29	4.20		The eastern one of the three
18ε	6♒29	−17;08	4.17		The star near the gills
14μ	27♑14	−19;55	4.49		The star in the southern spinal fin
ζ	4♒45	−15;26	6.43		The eastern one of the two in the belly
16λ	0♒34	−15;33	5.43		The western one of these
12η	27♑27	−15;06	5.42		The eastern one of the three in the northern fin
10θ	23♑48	−16;22	5.01		The middle one of these

CAT.	LONG.	LAT.	MAG.	NAME	DESCRIPTION
9 ι	22♑23	−18;12	4.33		The western one of the three
γ‡	23♑07	−22;57	5.45		The star at the tip of the tail
α**	10♑49	−15;14	4.92		The western one of the three bright stars west of the Fish
γ**	13♑38	−14;27	4.66		The middle one of these
ε**	17♑05	−15;27	4.70		The eastern one of the three
BSC 8076	14♑40	−14;51	5.18		The dim star west of this
BSC 8110	17♑31	−10;51	5.40		The southern star of the remaining two to the north
24A⁺	17♑04	−7;55	4.49		The northern one of these

*XXI;   **Microscopii;   ⁺Capricorni;   ‡Gruis

APPENDIX II

NEWTON'S CHRONOLOGY

ISAAC NEWTON DEVOTED some thirty or forty years of his life to the pursuit of the idea of relating historical events to the location of the equinoxes in the zodiac.[1] He intended to revise world chronology, to establish chronology on an astronomical basis, by linking the occurrence of historical events to the position of the equinoxes in the zodiac. It seems that he conceived of parallel histories in the heavens and on earth in which terrestrial events could be related to astronomical occurrences and vice versa. He endeavoured to establish this relationship on a scientific basis. From his astronomical system he determined the rate of precession to be 50" yearly, or 1° in 72 years.[2] By knowing the position of the equinoxes in the zodiac for a given year, he could calculate back to determine its location in the zodiac for past historical events.

> With equinoctial precession as an instrument any event in the past could be dated with certainty, provided that an ancient record could be found indicating the position of the sun at time of equinox relative to the fixed stars.... The aim of *The Chronology of Ancient Kingdoms Amended* (published posthumously in 1728) was the establishment of a relationship between the observed movement of the earth with respect to the fixed stars and ancient political events, so that the past might be 'predicted' backwards, so to speak.[3]

Applying this method, Newton calculated that the expedition of the Argonauts took place in 939 BC, when the equinoxes lay in the middle of the celestial signs of Aries and Libra (Chelae), i.e. when the vernal point was at 15° Aries.[4] He wrote further that:

> Thales revised Astronomy ... in his youth the Equinoxes were passing out of the 12th into ye 11th degrees of ye signes or Asterisms of Aries and Chelae ... Meton

---

1. Manuel, *Isaac Newton Historian*, p 17.
2. Cajori, *Newton's Mathematical Principles*, p 490.
3. Manuel, *Isaac Newton Historian*, p 68.
4. Ibid., p 73.

and Euctemon observed the summer solstice anno J. Per 4282 at wch time the cardinal points were passing out of the 9th into the 8th degree of the signes. . . .[1]

This latter observation corresponds to the year −431, since anno J. Per 4282 signifies the 4282nd year since the beginning of the 'julian period' with epoch − 4712.[2] Newton's sources for this are probably Ptolemy and Columella. From Ptolemy it can be deduced that Meton and Euctemon observed the summer solstice in −431,[3] and Columella states that Meton placed the equinoxes at 8° Aries−8° Libra, and the solstices at 8° Cancer−8° Sagittarius, respectively.[4]

Newton's chronological scheme is based on the sidereal zodiac, i.e. the division of the fixed-star zodiac into twelve 30° signs. His conception of the sidereal zodiac was obviously derived from Greek references to it mentioned above, since the Babylonian sidereal zodiac was not recovered until the end of the nineteenth century.[5] As the Greeks apparently had no precise definition of the sidereal zodiac, Newton depended upon the few references available to him, e.g. that of Columella, for his knowledge of the sidereal zodiac. It was not until cuneiform sources yielded information about the Babylonian prototype that an accurate definition of the sidereal zodiac could be formulated. The Babylonian sidereal zodiac is defined in Appendix I through a reconstruction of Ptolemy's star catalogue in which stellar longitudes are given in sidereal coordinates.

The Babylonian zodiac was transmitted to the Greeks, and Newton's knowledge of it was second-hand, derived from Greek literary sources. That his sources did not define the sidereal zodiac accurately is apparent from a consideration of the results which Newton obtained from applying his chronological scheme. Firstly, however, the following statement of Hipparchus, recorded by Ptolemy, allows an independent inquiry into the chronological validity of the Babylonian zodiac:

> For Hipparchus too, in his work *On the Displacement of the Solstitial and Equinoctial Points*, adducing lunar eclipses from among those accurately observed by himself, and from those observed earlier by Timocharis, computes that the distance by which Spica is in advance of the autumnal [equinoctial] point is about 6° in his own time, but was about 8° in Timocharis' time.[6]

---

1. Ibid., p75.
2. The 'julian period' (comprising 7980 years) was introduced in the sixteenth century by Scaliger in *De emendatione temporum*. Scaliger's innovation is a technical device for counting days, called 'julian days', with −4712 as epoch year. Cf. Neugebauer, HAMA III, pp1061 ff.
3. Ptolemy, *Almagest* III, 1 (trsl. Toomer, p138).
4. Columella, *De re rustica* IX, 14 (trsl. Ash-Forster-Heffner II, pp487/489).
5. By Epping, cf. Epping-Strassmaier (1893) p169, n1; cf. also Kugler, *Die Babylonische Mondrechnung*, pp73–74 and SSB I, pp27–31.
6. Ptolemy, *Almagest* VII, 2 (trsl. Toomer, p327).

As discussed in Chapter 2, the lunar eclipse observed by Timocharis is probably that which occurred at the vernal equinox in −283, whilst that observed by Hipparchus probably took place at the vernal equinox in −134. Since Spica's longitude in the Babylonian zodiac is 29° Virgo (Appendix I), Hipparchus' statement recorded by Ptolemy implies that in −283 the vernal equinox was located at 7° Aries and at 5° Aries in −134. This conforms with the definition of the Babylonian zodiac as given in Appendix I, which assumes that in −100 the longitude of the vernal point was 4°27′ Aries.[1] Using Newton's coefficient of precession, 1° in 72 years, the vernal equinox was at 0° Aries in AD 220, at 5° Aries in −140, and at 7° Aries in −284.[2] This result is in complete accordance with Hipparchus' statement above. Since Hipparchus is the most reliable Greek astronomical source prior to Ptolemy, this result establishes the chronological validity of the Babylonian zodiac in the sense of Newton's aim of dating historical events by locating the vernal point at specific degrees in the sidereal zodiac going back in history.

How does Newton's reckoning quoted above that the vernal point was at 15° Aries in −938 compare with the foregoing result? Computing back from AD 220, allowing a precession rate of 1° in 72 years, the vernal point retrogressed 15° in 1080 years, i.e. in the framework of the Babylonian sidereal zodiac as reconstructed in Appendix I the vernal point was at 15° Aries in −860 and at 16° Aries in −932. Hence there is a difference of about 1° in the sidereal zodiac used by Newton and that of the Babylonians as reconstructed in Appendix I, since Newton's chronology placed the vernal point at 15° Aries in −938. This difference leads to a chronological displacement of about 72 years between Newton's chronology and one based on the reconstructed Babylonian sidereal zodiac.

This discrepancy does not invalidate the principle of Newton's chronological scheme. It merely shows that Newton lacked a precise definition of the sidereal zodiac such as that given in Appendix I. The knowledge of the sidereal zodiac that he acquired from Greek sources was inaccurate, leading to a difference of about 1° from the Babylonian sidereal zodiac. How did this difference arise?

In Newton's *Chronology of Ancient Kingdoms Amended* he deals with the question of the definition of 15° Aries in relation to the fixed stars.[3] He considered 15° Aries to be pivotal on account of Hipparchus' reference to Eudoxus (fourth century BC) that he

---

1. The location of the vernal point at 4°27′ Aries in -100 agrees closely with Huber (1958), who locates the vernal point in the Babylonian zodiac at 4°28′ Aries for the epoch −100.
2. Appendix I introduction.
3. Newton, *The Chronology of Ancient Kingdoms Amended*, pp 86–91.

[Drew] the Colure of the Solstices through ... the middle of Cancer ... and through the middle of Capricorn; and that he drew the Equinoctial Colure through ... the middle of Chelae (Libra), and through ... the back of Aries ....[1]

Newton combined this information with his belief that Eudoxus' sphere was the same as that drawn up by Chiron at the time of the Argonautic expedition.

> 939 BC. The ship Argo is built.... Chiron, who was born in the Golden Age, forms the Constellations for the use of the Argonauts; and places the Solstitial and Equinoctial Points in the fifteenth degrees or middles of the Constellations of Cancer, Chelae, Capricorn, and Aries.... Meton in the year Nabonassar 316, observed the Summer Solstice in the eighth degree of Cancer, and therefore the Solstice had then gone back seven degrees. It goes back one degree in about seventy two years, and seven degrees in about 504 years. Count these years back from the year Nabonassar 316, and they will place the Argonautic expedition about 936 years before Christ.[2]

Newton believed that the primitive sphere of the Greeks with twelve equal constellations each 30° long was drawn up by Chiron in 939 BC when according to his calculation the vernal point was located at 15° Aries. However, as discussed in Chapter 5, the first use of sidereal coordinates in this way was the Babylonian System A in which the vernal point was located at 10° Aries, and the first rudimentary form of System A stems from the fifth century BC, nearly five hundred years after the date computed by Newton.

In 1689 Newton occupied himself with the problem of determining where 15° Aries stood in relation to the vernal point at his time, so that he could calculate back from 1689 to determine the date at which he believed Chiron had "formed the Constellations." According to Newton's identification of 15° Aries in the constellation of Aries, he determined first that the difference between 15° Aries and the vernal point in 1689 was 36°44',[3] and then subsequently he revised this to 36°29'.[4] Using a rate of precession of 1° in 72 years, he found the length of time taken for the vernal point to regress through 36°29' to be 2627 years. Going back 2627 years from 1689, he arrived at –938, i.e. 939 BC.[5]

---

1. Ibid., p83. As discussed in chap. 5, the placing by Eudoxus of the equinoxes and solstices in the middle of the signs is reminiscent of the schematic solar calendar from the tablet MUL.APIN (687 BC) in which the 15th day of the 1st, 4th, 7th, and 10th solar months coincides with the equinoxes and solstices (Figure 11). It is in his *Commentary to Aratus* (ed. Manitius, p48, 8) that Hipparchus informs us that Eudoxus placed the solstices and equinoxes at 15° of their respective signs.

2. Ibid., pp25–26. Since Nabonassar 316 = 432 BC, going back a further 504 years leads to 936 BC. Newton subsequently revised this date to 939 BC.

3. Ibid., p86.

4. Ibid., p91.

5. 36°29' = 36.4833° x 72 = 2627 subtracted from 1689 = –938.

When Fréret published his French translation of Newton's *Abstract of Chronology*, he included a confutation in which he misunderstood Newton's identification of 15° Aries.[1] Newton wrote in reply:

> I follow Eudoxus, and, by doing so, place the Equinoctial Colure about 7°36' from the first Star of Aries. But the Observator [Fréret] represents that I place it fifteen degrees from the first Star of Aries, and thence deduces that I should have made the Argonautic Expedition 532 years earlier than I do.[2]

Thus Newton identified 15° Aries to be in the center of the constellation of Aries, 7°36' east of Mesartim (γ Arietis), the first star of Aries. In the star catalogue of Copernicus, Mesartim is the prime fiducial star from which the longitudes of all the other stars are measured.[3] Like the reconstructed Babylonian star catalogue in Appendix I, Copernicus' star catalogue is sidereal, the difference being that he took Mesartim (0° Aries) as his reference star instead of Aldebaran (15° Taurus), the prime fiducial star chosen by the Babylonians. Fréret wrongly assumed that Newton was using the same principle as Copernicus, in which 15° Aries could be interpreted as 15° east of Mesartim, whereas in 1689 Newton actually had specified that 15° Aries was 7°36' east of Mesartim. Based on this false assumption, Fréret concluded that 15° Aries lay 7°24' further to the east than the location determined by Newton. This wrong assumption led Fréret to conclude that the vernal point had been located at 15° Aries in 1471 BC, some 532 years before 939 BC, the date arrived at by Newton.[4]

The real basis for Newton's computation, however, is Columella's statement concerning the observation that Meton in 432 BC *observed the summer solstice in the eighth degree of Cancer*.[5] From this Newton concluded that the vernal point was at 8° Aries in 432 BC. Going back a further 7° to 15° Aries at a precession rate of 1° in 72 years, i.e. going back 7 x 72 = 504 years from 432 BC, leads back to 936 BC, which—as indicated above—Newton subsequently revised to 939 BC This is how Newton arrived at his date of 939 BC for the date of the location of the vernal point at 15° Aries.

---

1. Nicolas Fréret, *Abrégé de Chronologie de M. le Chevalier Newton, fait per lui meme, et traduit sur le Manuscript Anglais*. The best English version of Newton's *Abstract of Chronology* is *A Short Chronicle from the First Memory of Things in Europe to the Conquest of Persia by Alexander the Great* published as pages 1 to 42 in *The Chronology of Ancient Kingdoms Amended*, first published in 1728. The French pirated edition translated by Fréret from a manuscript copy made without Newton's permission appeared in 1725 as a supplement to a French version of the seven volumes of Dean Humphrey Prideaux's *History of the Jews and Neighbouring Nations, from the Declension of the Kingdom of Israel and Judah to the time of Christ*.
2. Newton (1725), p318.
3. Copernicus, *De revolutionibus orbium coelestium* II, 14 (trsl. Rosen, pp83–84 and p97).
4. 7°24' = 7.4 x 72 = 532.8 years, added to 939 BC leads back to 1472 BC.
5. Newton, *The Chronology of Ancient Kingdoms Amended*, pp25–26.

As discussed in Chapter 3, the time of the summer solstice in 432 BC was around 10:30 a.m. (local time in Athens) on 28th June, when the sun was at 9° Cancer in the Babylonian sidereal zodiac as reconstructed in Appendix I. However, owing to a one day error in his determination of the day of the solstice, which he determined to be on 27th June in that year, even though he determined the sidereal longitude of the solstitial point incorrectly (at 8° Cancer instead of 9° Cancer), Meton recorded the longitude of the sun *correctly*, since the sun was actually at 8° Cancer one day earlier, on 27th June 432 BC. This finding indicates that Meton was evidently using the Babylonian sidereal zodiac but that he made a 1° error in his specification of the location of the summer solstitial point in 432 BC because he thought that the summer solstice occurred on 27th June when actually it occurred one day later. It is this 1° error,[1] taken over by Newton, that accounts for the 1° discrepancy between Newton's chronology and that given by the regression of the vernal point through the Babylonian sidereal zodiac, in which the vernal point was located at 15° Aries some 1080 years before AD 220, i.e. in –860 (861 BC). and not in 939 BC as computed by Newton.

Since Newton's time a new chapter in historiography has unfolded. Following the excavation of hundreds of thousands of clay tablets from the ruins of Babylon and Uruk, much of Mesopotamian history has been reconstructed. Similarly, the translation of Egyptian, Demotic, and Greek papyri, and the decipherment of texts inscribed on ancient Egyptian monuments, coffin lids, etc., has enabled a reliable reconstruction of Egyptian history to be established. It is possible that Newton's chronological scheme, based on a secure definition of the sidereal zodiac, may be applicable to ancient history, as he envisaged.

---

1. As discussed in Chapter 3, it is implicit in Ptolemy's statement "it is reasonable to reckon the beginnings of the signs also from the equinoxes and solstices" (*Tetrabiblos* I, 22; trsl. Robbins, p109) that he considered the location of the vernal point to be at 0° Aries in his day. As he wrote the *Tetrabiblos* after the *Almagest*, and as the epoch date for his star catalogue in the *Almagest* is AD 138, possibly he wrote the *Tetrabiblos* around AD 148, when the vernal point was located at 1° Aries (Appendix I), signifying a 1° difference, just as Meton some 580 years before him, through having made a one day error in determining the day of the summer solstice, was also differing by 1° with regard to the location of the summer solstitial point (and thus of the vernal point) in the Babylonian sidereal zodiac. In Ptolemy's case it is not at all an error but simply the circumstance of his using the tropical zodiac (vernal point at 0° Aries), whereas the vernal point was at 1° Aries in the Babylonian sidereal zodiac at that time. On the other hand, in compiling his catalogue of stars Ptolemy made a systematic error amounting to 1°14' in his location of the vernal point, signifying that a correct epoch date for his catalogue would have been around AD 48, some 89 years earlier than the actual epoch date of AD 137/138, since his star catalogue would have been correct with the vernal point displaced by 1°14' considering that it took the vernal point around 89 years to retrogress through 1°14' between AD 48 and AD 137; cf. Neugebauer, HAMA I, p284.

[The] thousands upon thousands of newly unearthed inscriptions on common stone and marble, the papyri and the clay tablets, have opened up an ancient world which men of his (i.e. Newton's) age, restricted to fragments of literature, never envisioned.... Dating events through computations of the equinoctial precession ... remains sound in theory and could be effectively applied if only ancient astronomical observations, translated into modern star catalogues, would come to light.[1]

It is unlikely, however, that ancient observations such as those of Timocharis and Hipparchus, in which the equinoctial or tropical points are related to fixed stars, will come to light. The inherent difficulty in observing the stellar location of the equinoxes or solstices means that no accurate determination of their sidereal location was possible before a certain level of observational precision was attained in astronomy. In Greek astronomy there were endeavors, as discussed above (Chapter 3), to observe the sidereal location of the equinoxes and the solstices, e.g. by Timocharis and Hipparchus, and even earlier, according to Columella, by Meton. But it was not until Hipparchus that the equinoctial and tropical points became determined with sufficient accuracy to enable them to be used in order to redefine the zodiac.[2] In Babylonian astronomy the equinoxes and solstices were observed from very early times, pre-seventh century BC, by the use of a 'gnomon', a vertical rod whose shadow length was observed to tell the time of day.[3] Later, Babylonian astronomers employed simple calendrical schemes for determining the lunar month and day on which the solstices and equinoxes would occur.[4] However, these early schemes were not concerned with finding the location of the tropical and equinoctial points in relation to the fixed stars. It was only after the sixth century BC, when

---

1. Manuel, *Isaac Newton Historian*, p191.
2. The redefinition of the zodiac that seems to have occurred with Hipparchus, if indeed it was Hipparchus who introduced the tropical zodiac into astronomy (Chapter 3), is the principle of measuring from the solstitial and equinoctial points (tropical zodiac) rather than in relation to stars in the zodiacal belt (sidereal zodiac) as the Babylonians did. This step from the Babylonian sidereal zodiac to the Greek tropical zodiac was only possible when the locations of the solstitial and equinoctial points could be determined relatively accurately. Neugebauer, HAMA I, p284 points out that both Hipparchus and Ptolemy were somewhat inaccurate in their estimation of the location of the vernal point, Ptolemy more so than Hipparchus. The star catalogue of Hipparchus is reckoned to have had an epoch date of -127 but his stellar longitudes would have been correct about 11 years earlier, in -138. This is actually only a small error (0°09') in determining the location of the vernal point. However, Ptolemy's star catalogue had an epoch date of AD 137/138 and his stellar longitudes (measured from the vernal point) would have been correct about 89 years earlier, in AD 48, indicating an error of 1°14' in his specification of the location of the vernal point (cf. also n1 on previous page).
3. Neugebauer, HAMA I, pp544 f.
4. Ibid., pp357–363, and pp542 ff; the earliest extant solstice-equinox text refers to the years −615 to −587 (HAMA II, p542). Cf. also Hunger-Pingree (1999), pp151–152.

the coordinate system of the Babylonian sidereal zodiac was introduced, that the vernal point was given a sidereal longitude[1]—for example, 10° Aries in System A and 8° Aries in System B of Babylonian astronomy. The earliest period, then, at which observations or records of the stellar location of the equinoxes or solstices may be expected is from around the fifth century BC onwards.

---

1. Neugebauer, HAMA II, p594.

AFTERWORD

THIS THESIS ON THE DEFINITION and transmission of the zodiac is concerned with the original specification of the zodiac, i.e. the first scientific definition of the zodiac.

According to this original definition the zodiac is defined by the two first magnitude stars Aldebaran and Antares in such a way that each is located exactly at the midpoint (15°) of their respective sign, Taurus and Scorpio. Thereby these two stars define the central axis of the zodiac, which was the primary zodiacal reference axis for all other stars. On the basis of observation and measurement in relation to this reference axis, the brighter stars near the ecliptic belonging to the twelve zodiacal constellations were assigned longitudes in the zodiac, for example the stars α and β Librae were assigned the longitudes 20° and 25° Libra. Thus the first astronomical coordinate system came into being, through which the positions of the stars and planets along the ecliptic could be determined between 0° and 30° within the twelve zodiacal signs.

This coordinate system of twelve zodiacal signs, i.e. twelve equal-length zodiacal constellations each 30° long, emerged in Babylonian astronomy during the fifth century BC and was transmitted from Babylon to Greece, Hellenistic Egypt, Rome, and India. It had a forerunner in the schematic solar calendar of MUL.APIN (seventh century BC), which comprised a division of the year into twelve solar months relating to the sun's movement in declination.[1] This solar calendar was possibly the forerunner of that of the Greek astronomer Euctemon (fifth century BC). Euctemon's solar calendar went through a metamorphosis—probably through Hipparchus (second century BC)—to become the tropical zodiac, one of the standard coordinate systems of astronomy.[2]

---

1. MUL.APIN, a series of astronomical texts on two tablets, means "the plow star"—named according to the initial words on the first tablet. MUL.APIN stems from the seventh century BC but contains observations that were made much earlier. The solar calendar of MUL.APIN consists of a schematic year of twelve months, each 30 days long, such that the equinoxes and solstices fall on the 15th day of the months I, IV, VII, and X. A correspondence is evident between the solar calendar and the zodiac: the zodiac consists of twelve signs each 30° long, corresponding to the solar calendar of twelve months each 30 days in length.

2. In the MUL.APIN solar calendar the equinoxes are placed on the 15th day of the months I and VII and the solstices on the 15th day of the months IV and X, whereas in Euctemon's solar

A list of 17 constellations, some of whose stars lie within the moon's path on or close to the ecliptic, is also given in MUL.APIN. This too, since all twelve zodiacal constellations are named in this list, was a forerunner of the zodiac.

Between the fifth and the third centuries BC a development occurred from the early MUL.APIN level of Babylonian astronomy to the highly developed mathematical astronomy of Systems A and B based on computations using the ecliptic coordinate system of the sidereal zodiac as defined above in relation to the stars.[1] This new level of astronomy relied on the one hand on the ecliptic coordinate system of the Babylonian sidereal zodiac and on the other hand on the availability of records of astronomical observations collected over an extended period of time.

It is unknown through whom the innovation leading to the introduction of the zodiacal coordinate system into Babylonian astronomy took place. It was made possible by the observation of the heavens as a systematic program carried out over hundreds of years, and it is an astonishing fact of the history of

---

calendar the equinoxes are located on the *first* day of the months I and VII and the solstices on the *first* day of the months IV and X. This is a new principle in relation to the Babylonian solar calendar. This new calendar principle introduced by Euctemon is mirrored in the definition of the tropical zodiac, in which the vernal and autumnal points are placed at 0° Aries and 0° Libra and the summer and winter solstitial points are located at 0° Cancer and 0° Capricorn. As discussed in note 1 (p206), it was probably Hipparchus (second century BC) who brought about the transformation of Euctemon's *temporal* solar calendar into the *spatial* coordinate system of the tropical zodiac. In contrast, according to Hipparchus the Greek astronomer Eudoxus (fourth century BC) defined the vernal and autumnal points to be at 15° Aries and 15° Libra and the summer and winter solstitial points to be at 15° Cancer and 15° Capricorn. This definition by Eudoxus corresponds exactly to the solar calendar from MUL.APIN (see footnote 1 below).

1. The ecliptic is the apparent path of the sun through the center of the zodiac. The original zodiac with twelve signs, each 30° long, specified by the Babylonians—in which Aldebaran is located at 15° Taurus and Antares at 15° Scorpio—is called the *sidereal zodiac*, since it is defined in relation to the stars (sidereal means "of the stars"), in order to distinguish it from the *tropical zodiac*, which is defined in relation to the vernal point. The sidereal zodiac and the tropical zodiac both comprise the same ecliptic coordinate system of twelve signs, each 30° long. However, whereas the sidereal zodiac is defined in relation to the stars, the tropical zodiac is defined in relation to the vernal point, which is equated with 0° Aries as the zero point of the tropical zodiac. Note that if there was a perfect correspondence between MUL.APIN's solar calendar and the zodiac (see footnote 1 on previous page), the vernal point would have to be located at 15° Aries, since Aries as the first sign of the zodiac corresponds to month I and in the Babylonian solar calendar the vernal equinox was placed on the 15th day of month I (as with Eudoxus' definition, see footnote 2 above). However, in System A of Babylonian astronomy the vernal point was located at 10° Aries and in System B at 8° Aries. From this some researchers concluded that the Babylonians already knew about the slow movement of the vernal point ("precession of the equinoxes") backwards through the constellations—one sign (30°) in 2160 years, i.e. 1° in 72 years—and that they observed the position of the vernal point in the constellation of Aries and thus determined its location at that time to be 10° Aries (System A) or 8° Aries (System B).

science that this systematic observation of the night sky was executed by Babylonian astronomers over many centuries.

In the last analysis this thesis pays tribute to those early astronomers who kept records of their observations and were able on this basis to arrive at the innovation of the ecliptic coordinate system of the zodiac, which was central to the whole subsequent development of astronomy as a science.

The original contribution of this thesis to the history of astronomy is the uncovering of the *intrinsic definition* of the Babylonian sidereal zodiac. This definition is intrinsic, because it is nowhere explicitly stated in the available cuneiform sources. Nevertheless the placing of Aldebaran at 15° Taurus and Antares at 15° Scorpio is the foundation for the definition of the zodiac.

Observation of the starry heavens reveals that the twelve zodiacal constellations (with the exception of Libra) are distributed such that their centers lie approximately 30 degrees from one another, and in the case of the constellations of Taurus and Scorpio their centers are marked by the bright stars Aldebaran and Antares. That which I have called the *intrinsic definition* of the zodiac, comprising the central content of this thesis, probably derives from the simple observation of the distribution of the circle of the twelve zodiacal constellations (as a division into twelve 30°-sectors or signs) around the celestial sphere.

This original definition of the zodiac of the Babylonians in the fifth century BC was a significant event in the history of astronomy as it heralded the beginning of the development of mathematical astronomy. Babylonian mathematical astronomy reached a highpoint during the Seleucid era (third and second centuries BC). However, although the Greek astronomer Hipparchus adopted some Babylonian astronomical parameters, he effectively redefined the zodiac when, rather than utilizing the Babylonian sidereal zodiac, he used instead the tropical zodiac, if indeed he really did utilize the tropical zodiac.[1] Although the sidereal zodiac was transmitted from Babylon to Greece, Hellenistic Egypt, Rome, and India, it was replaced in Greek astronomy by the tropical zodiac. Thus the tropical zodiac is the standard astronomical coordinate system used in the second century AD by Ptolemy in the *Almagest*.

In addition to exploring in detail the original definition of the zodiac, a second original contribution of this thesis is the uncovering of the line of descent from the schematic solar calendar of the Babylonians from MUL.APIN to the tropical zodiac of Greek astronomy (Hipparchus and Ptolemy) via the solar calendar of Euctemon. Essential to understanding this line of descent is the insight that the Babylonian solar calendar and Euctemon's solar calendar are both an expression of the sun's movement in declination.

The tropical zodiac, then, as a spatial projection of Euctemon's solar calendar, is actually based on the sun's movement in declination, which is intrinsically independent of the sun's motion in longitude through the sidereal zodiac. Thus midsummer always occurs when the sun attains its maximum declination. However, the midsummer sun for Meton and Euctemon (432 BC) was in the constellation of Cancer (Meton specified it to be at 8° Cancer whereas it was actually at 9° Cancer in the Babylonian sidereal zodiac at that time). At the present time, however, on account of precession the midsummer sun is located in the constellation of Gemini (currently at 5° Gemini in the Babylonian sidereal zodiac). From this it is readily apparent that the two coordinate systems—that of the Babylonians (the sidereal zodiac) and the one favored in Greek astronomy (the tropical zodiac)—are intrinsically independent. To what extent this was grasped by Ptolemy—in particular regarding the long-term consequence of adopting the tropical zodiac as the standard astronomical frame of reference, thus signifying the gradual spatial displacement between the tropical and sidereal zodiacs on account of precession—is unknown. Although Ptolemy lacked the conceptual framework and the vantage point of modern astronomy, it was theoretically possible for him to anticipate the future spatial dislocation of the tropical zodiac in relation to the background of the stellar constellations that is apparent now that the vernal point is located in the region of the Western Fish of the constellation of Pisces, i.e. now that the displacement between the tropical and sidereal zodiacs amounts to some 25° (almost one complete zodiacal sign).

1. It is generally assumed that Hipparchus was the first to use the tropical zodiac. In his *Commentary on the Phaenomena of Aratus and Eudoxus* Hipparchus used a variety of coordinate systems, but he did not use the ecliptic coordinate system of the tropical zodiac in this work. The evidence that Hipparchus used the tropical zodiac as a coordinate system is inconclusive. The most important evidence is provided by Ptolemy's remarks in *Almagest VII*, 2 concerning Hipparchus' measurements of the longitudes of the stars Regulus and Spica. However, the possibility has to be considered that Ptolemy himself converted Hipparchus' measurements into the ecliptic coordinate system of the tropical zodiac. There is also the statement by Columella (first century AD) in *De re rustica IX*, 14 that Hipparchus placed the solstices and equinoxes in the first degrees (that is, at 0°) of the signs of the zodiac, i.e. that he placed the vernal point at 0° Aries, which defines the tropical zodiac. It is an assumption based on such remarks made by Ptolemy and Columella, rather than evidence from Hipparchus himself, that he, after having discovered the precession of the equinoxes, realized the importance of the ecliptic as a coordinate system and then compiled a star catalog using the ecliptic coordinate system of the tropical zodiac—presumably in the (no longer extant) star catalog said to have been compiled by Hipparchus as mentioned by Pliny in *Natural History II*, 95 and attested indirectly by Ptolemy in the *Almagest* with his references to Hipparchus' interest in observing the positions of the fixed stars. It would seem that Hipparchus probably introduced the tropical zodiac into Greek astronomy. What is definitely known about the tropical zodiac is that it is the standard ecliptic coordinate system that was used by Ptolemy in the second century AD and which—following Ptolemy's example—was used in astronomy thereafter.

As the tropical zodiac is still used as a coordinate system in modern astronomy, the uncovering of the line of descent from the solar calendar of the Babylonians to the tropical zodiac reveals the indebtedness of modern astronomy to its origins in Babylonian astronomy. Even if Euctemon arrived at his solar calendar (underlying the tropical zodiac) independently of the Babylonian solar calendar, the line of descent *as a principle in the history of astronomical ideas* is nevertheless apparent, i.e. that Babylonian astronomers were the first to conceive of expressing the sun's yearly motion in declination in the form of a solar calendar. We shall probably never know if Euctemon derived his solar calendar from that of the Babylonians, but we can at least honor the achievement of Babylonian astronomers as being the first to conceive of such a solar calendar, just as this thesis honors their great achievement in defining the original zodiac: the sidereal zodiac.

BIBLIOGRAPHY

Aaboe, Asger. "On the Babylonian Origin of Some Hipparchian Parameters," *Centaurus* 4 (1955–56), pp 122–125.

———. "Scientific astronomy in antiquity," *Philosophical Transactions of the Royal Society of London* A 276 (1974), pp 21–42.

Aaboe-Sachs: Aaboe, Asger—Sachs, Abraham. "Two Lunar Texts of the Achaemenid Period from Babylon," *Centaurus* 14 (1969), pp 1–22.

Albiruni. *The Chronology of Ancient Nations*, trsl. C.E. Sachau (1879). London.

Allen, Richard Hinckley. *Star Names, Their Lore and Meaning.* (1963 repr. from 1899). New York.

Almagest: see Ptolemy.

Arthava-veda Samhita, trsl. W.D. Whitney (1905). Cambridge, Mass.

The Astronomical Ephemeris. (yearly). London and Washington.

Autolycus. Greek text with English trsl.: *The Books of Autolycus. On a Moving Sphere and on Risings and Settings*, ed. and trsl. F. Bruin and A. Vondjidis (1971). Beirut.

Barton and Barton: S.G. Barton and W.H. Barton. *A Guide to the Constellations* (1943). New York.

Bhishma Purva: see *Mahabharata*, vol. 6.

Bidez, Joseph. *Bérose et la grande année* (1904). Bruxelles.

Billard, Roger. *L'astronomie indienne* (1971). Paris.

Biot, Édouard. "Catalogue des étoiles extraordinaires observées en Chine depuis les temps anciens jusqu'à l'an 1203 de notre ère. Extrait du livre 294 de la grande collection de Ma-touan-lin," *Connaissance de Temps pour l'an 1846* (1843), pp 60–68. Paris.

Boll, Franz. *Sphaera* (1903). Leipzig.

Bouché-Leclerq, Antoine. *L'astrologie grecque* (1963 repr. from 1899). Paris.

Brennand, W. *Hindu Astronomy* (1896). London.

Britton, John P. "Scientific Astronomy in Pre-Seleucid Babylon," *Die Rolle der Astronomie in den Kulturen Mesopotamiens* (1993), pp 61–76. Graz, Austria.

———. "Lunar Anomaly in Babylonian Astronomy," *Ancient Astronomy and Celestial Divination*, ed. N.M. Swerdlow (1999). Cambridge, Mass.

Britton-Walker: Britton, John P.—Walker, Christopher. "Astronomy and Astrology in Mesopotamia," *Astronomy before the Telescope*, ed. C. Walker (1996), pp 45–56. London.

Cajori, Florian. *Sir Isaac Newton's Mathematical Principles* (1934). London.

Cleomedes, *De motu circulari corporum caelestium libri duo*, ed. H. Ziegler (1891). Teubner. Leipzig.

Colebrooke, H.T. *Miscellaneous Essays* (2 vols., 1837). London.

Columella, Lucius Junius Moderatus. *De re rustica*: Latin text with English trsl.: *On Agriculture*, trsl. H.B. Ash, E.S. Forster, and E.H. Heffner (3 vols., 1941, 1954, 1955). Loeb Classical Library.

Copernicus, Nicolaus. *De revolutionibus orbium coelestium*. Facsimile reprint of the first edition of 1543 (1965). New York. English trsl.: *On the Revolutions*, trsl. Edward Rosen, *Nicholas Copernicus*:

———. Vol. I, *Complete Works* (2 vols., 1992). Baltimore.

Cramer, Frederick H. *Astrology in Roman Law and Politics*. (*Memoirs of the American Philosophical Society*, vol. 37, 1954). Philadelphia.

Danckwortt, Albert. Danckwortt's catalogue: *Vierteljahrschrift der Astronomischen Gesellschaft* 16 (1881). Leipzig.

Delporte, E. *Atlas céleste* (1930). Cambridge.

———. *Délimitation scientifique des constellations* (1930). Cambridge.

Dictionary of Scientific Biography (1970 ff). New York.

Diogenes Laertius. *Lives: Lives of Eminent Philosophers*, Greek text with English trsl., R.D. Hicks (2 vols., 1925). Loeb Classical Library.

Duhem, Pierre. *Le Système du Monde* (10 vols., 1913–59). Paris.

Dumont, Paul-Emile. "The Istis to the Nakṣatras (or Oblations to the Lunar Mansions) in the Taittiriya-Brahmana," *Proceedings of the American Philosophical Society* 98 (1954), pp 204–223.

Epping, Joseph. *Astronomisches aus Babylon: Stimmen aus Maria Laach, Ergänzungsheft* 44 (1889). Freiburg i. B.

Epping-Strassmaier: Epping, Joseph–Strassmaier, Johann N. "Der Saros-Canon der Babylonier," *Zeitschrift für Assyriologie* 8 (1893), pp 149–178.

Evans, James. *The History and Practice of Ancient Astronomy* (1998). Oxford.

Euclid. *Phaenomena: Euclidis opera omnia* viii, ed. J.L. Heiberg and H. Menge (1916), pp 1–156. Leipzig. German trsl.: *Euclids Phaenomena*, trsl. A. Nokk (1850). Freiburg i. Br.

Fagan, Cyril. *Zodiacs Old and New* (1951). London.

Gleadow, Rupert. *The Origin of the Zodiac* (1968). London.

Goldstein-Bowen: Goldstein, Bernard R.—Bowen, Alan C. "On Early Hellenistic Astronomy: Timocharis and the First Callippic Calendar," *Centaurus* 32 (1989), pp 272–293.

Graßhoff, Gerd. *The History of Ptolemy's Star Catalogue* (1990). New York.

———. "Normal Star Observations in Late Astronomical Babylonian Diaries," *Ancient Astronomy and Celestial Divination*, ed. N.M. Swerdlow (1999), pp 97–147. Cambridge, Mass.

Gundel, Wilhelm. *Dekane und Dekansternbilder* (1936). Hamburg.

———. *Neue astrologische Texte des Hermes Trismegistos* (1978 repr. from 1936). Hildesheim, Germany.

Heath, Sir Thomas L. *A History of Greek Mathematics* (2 vols., 1931). London.

———. *Greek Astronomy* (1932). London.

Hephaestion of Thebes, *Apotelesmatica*, ed. D. Pingree, *Hephaestionis Thebani Apotelesmaticorum libri tres* (vol. I, 1973; vol. II, 1974). Teubner. Leipzig.

August Engelbrecht, *Hephaestion von Theben und sein astrologisches Compendium* (1887). Vienna.

Herschel, John F. *Outlines of Astronomy* (1849). London.

Hesiod. *Works and Days*. (*Hesiod, the Homeric Hymns and Homerica;* trsl. H.G. Evelyn White, 1936). Loeb Classical Library: London & Cambridge, Mass.

Hipparchus. *Commentary on the Phaenomena of Aratus and Eudoxus*, abbreviated: *Commentary to Aratus*: Greek text and German trsl.: *Hipparchi in Arati et Eudoxi Phenomena Commentarius*, ed. and trsl. C. Manitius (1894). Leipzig.

Huber, Peter. "Ueber den Nullpunkt der babylonischen Ekliptik," *Centaurus* 5 (1958), pp 192–208.

Hunger-Pingree: Hunger, Hermann—Pingree, David. MUL.APIN, *An Astronomical Compendium in Cuneiform* (*Archiv für Orientforschung, Beiheft* 24, 1989). Horn, Austria.

———. *Astral Sciences in Mesopotamia* (1999). Leiden-Boston-Cologne.

Joannis Stobaei. *Eclogarum Physicarum et Ethicarum*, ed. A. Meineke (2 vols., 1860, 1864). Leipzig.

Kane, Pāṇḍuraṅga Vāmana. *History of Dharmaśāstra* (5 vols., 1930–62). Poona, India.

Kennedy, Edward S. "Ramifications of the World-Year Concept in Islamic Astrology," *Ithaca. Actes du dixième congrès international d'histoire des sciences* (1964). Paris.

Kennedy-Pingree: Kennedy, E. S.—Pingree, David. *The Astrological History of Mashā'allāh* (1971). Cambridge, Mass.

Kennedy-van der Waerden: Kennedy, E.S.—van der Waerden, B.L. "The World-Year of the Persians," *Journal of the American Oriental Society* 83 (1963), pp 315–327.

Koch-Westerholz: Koch, J.—Westerholz, U. *Mesopotamian Astrology* (1995). Copenhagen.

Kollerstrom, Nicholas. "The Star Zodiac of Antiquity," *Culture and Cosmos* 1 (1997), pp 15–22.

———. "On the Measurement of Celestial Longitude in Antiquity," *Optics and Astronomy*, ed. G. Simon and S. Débarbat (2001), pp 145–159. Bruxelles.

Kudlek-Mickler: Kudlek, Manfred—Mickler, Erick H. *Solar and Lunar Eclipses of the Ancient Near East from 3000 BC to 0* (1971). Neunkirchen-Vluyn, Germany.

Kugler, Franz Zavier. *Die Babylonische Mondrechnung* (1900). Freiburg i. B., Germany.

———. SSB: *Sternkunde und Sterndienst in Babel* i (1907), ii (1909–1924). Münster.

Kunitzsch, Paul. *Arabische Sternnamen in Europa* (1959). Wiesbaden, Germany.

———. *Der Almagest. Die Syntaxis Mathematica des Claudius Ptolemäus in arabisch-lateinischer Üeberlieferung* (1974). Wiesbaden, Germany.

Langdon, Stefan. *Babylonian Menologies and the Semitic Calendars* (1935). London.

Lasserre, F. *Die Fragmente des Eudoxos von Knidos*. (1966). Berlin.

Lewis, David. "Extrinsic Properties," *Philosophical Studies* 44 (1983), pp 111–112.

MacKenzie, D.N. "Zoroastrian Astrology in the *Bundahišn*," *Bulletin of the School of Oriental and African Studies* 27 (1964), pp 511–529.

Mahabharata, trsl. M.N. Dutt (18 vols., 1895–1905). Calcutta.

Manilius. *Astronomica*, trsl. G.P. Goold (1977). Loeb Classical Library.

Manuel, Frank E. *Isaac Newton Historian* (1963). Cambridge.

Mercier, Raymond. "Studies in the Medieval Conception of Precession," *Archives internationales d'histoire des sciences* 26 (1976), pp 197–220, and 27 (1977), pp 33–71.

Needham, Joseph. *Science and Civilization in China* (5 vols., 1954 ff). Cambridge.

Neugebauer, Otto. ACT: *Astronomical Cuneiform Texts* (3 vols., 1955). London.

———. ESA: *The Exact Sciences in Antiquity* (1969). New York.

———. HAMA: *A History of Ancient Mathematical Astronomy* (3 vols., 1975). Berlin-Heidelberg-New York.

———. (1942): "Egyptian Planetary Texts," *Transactions of the American Philosophical Society* NS 32 (1942), pp 210–250.

———. (1943): "Demotic Horoscopes," *Journal of the American Oriental Society* 63 (1943), pp 115–127.

———. (1950): "The Alleged Babylonian discovery of the Precession of the Equinoxes," *Journal of the American Oriental Society* 70 (1950), pp 1–8.

———. (1959): "The Equivalence of Eccentric and Epicyclic Motion According to Apollonius," *Scripta Mathematica* 24 (1959), pp 5–21.

Neugebauer-Parker: Neugebauer, Otto—Parker, Richard A. EAT: *Egyptian Astronomical Texts* (3 vols., 1960, 1964, 1969). Providence.

Neugebauer-van Hoesen: Neugebauer, Otto—van Hoesen, H.B. *Greek Horoscopes* (*Memoirs of the American Philosophical Society*, vol. 48, 1987 repr. from 1959). Philadelphia.

Neugebauer, Paul Victor. Neugebauer's catalogue: *Sterntafeln von 4000 vor Chr. bis zur Gegenwart nebst Hilfsmitteln zur Berechnung von Sternpositionen zwischen 4000 vor Chr. und 3000 nach Chr.* (1912). Leipzig.

Newton, Isaac. "Remarks upon the Observations made upon a Chronological Index of Sir Isaac Newton, translated into French by the Observator, and Published at Paris," *Philosophical Transactions of the Royal Society* 33 (1725), pp 315–321.

_____. *The Chronology of Ancient Kingdoms Amended* (1728). London.

North, John D. *The Fontana History of Astronomy and Cosmology* (1994). London.

Peters-Knobel: Peters, Christian Heinrich Friedrich — Knobel, Edward Ball. *Ptolemy's Catalogue of Stars* (1915). Washington.

Piazzi, Giuseppe. Piazzi's catalogue: "Piazzi's Beobachtungen 1792–1813," *Kaiserlich-Königliche Sternwarte Annalen* Neue Folge. 4–12 (1821 ff). Vienna.

K. A. Pickering, "A Re-identification of some entries in the Ancient Star Catalog," *Dio* 12 (2002), pp 59–66; cf. also *Dio* 2 (1992), p 15.

Pillai, L. D. Swamikannu. *An Indian Ephemeris* (6 vols., 1922). Madras.

Pinches-Sachs: Pinches, T. G.-Strassmaier, J. N.–Sachs, A. J. *Late Babylonian Astronomical and Related Texts* (1955). Providence.

Pingree, David. *Census of the Exact Sciences in Sanskrit* (1970 ff). Philadelphia.

_____. (1963):"Astronomy and Astrology in India and Iran," *Isis* 54 (1963), pp 229-246

_____. (1968):*The Thousands of Abū Ma'shar*. London.

_____. (1973): "The Mesopotamian Origin of early Indian Mathematical Astronomy," *Journal for the History of Astronomy* 4 (1973), pp 1–12.

_____. (1976): "The Recovery of Early Greek Astronomy from India," *Journal for the History of Astronomy* 7 (1976), pp 109–123.

_____. (1978): *The Yavanajātaka of Sphujidhvaja*, ed. and trsl. D. Pingree (*Harvard Oriental Series* 48, 2 vols., 1978). Cambridge, Mass.

_____. (1989): "MUL.APIN and Vedic Astronomy," *DUMU-E_2-DUB-BA-A. Studies in Honor of Åke W. Sjöberg* (1989), pp 439–445. Philadelphia.

_____. See also Hunger-Pingree, Kennedy-Pingree, and Reiner-Pingree.

Pliny. NH: *Plinius. Naturalis historia*, ed. Jan-Mayhoff (5 vols., 1897–1933). Leipzig. English trsl.: *Pliny, Natural History*, trsl. H. Rackham et al. (10 vols., 1938–1962), Loeb Classical Library.

Porphyry. *Life of Pythagoras*, trsl. K. S. Guthrie (*The Pythagorean Sourcebook and Library*, 1987), pp 123–135. Grand Rapids, Michigan.

Powell, Robert. *The Zodiac: A Historical Survey* (1984). San Diego.

Powell–Treadgold: Powell, Robert–Treadgold, Peter. *The Sidereal Zodiac* (1985). Tempe, Arizona.

Pritchett-van der Waerden: Pritchett, W. Kendrick—van der Waerden, B. L. "Thucydidean Time-Reckoning and Euctemon's Seasonal Calendar," *Bulletin de Correspondance Hellénique* 85 (1961), pp 17–52.

Ptolemy: Claudius Ptolemaeus. *Almagest*: Greek text: *Claudii Ptolemaei opera quae extant omnia: Syntaxis Mathematica*, ed. J. L. Heiberg (1898, 1903). Leipzig. English trsl.: *Ptolemy: The Almagest*, trsl. R. C. Taliaferro, *Great Books of the Western World* 16 (1952), pp 1–478. Chicago.

_____. *Ptolemy's Almagest*, trsl. and annotated by G. J. Toomer (1984). New York-Berlin. Paperback edition by Princeton University Press (1998).

---. *Tetrabiblos*: Greek text with English trsl.: Ptolemy, *Tetrabiblos*, ed. and trsl. F.E. Robbins (1940). Loeb Classical Library.

Rehm, Albert. RE Par.: "Parapegma," *Pauly Wissowa. Realencyclopädie der classischen Altertumswissenschaft* 18, 4 col. 1295–1366.

---. "Griechische Kalender, III. Das Parapegma des Euktemon," *Sitzungsberichte Heidelberger Akademie der Wissenschaft* 3 (1913).

Reiner-Pingree: Reiner, Erica—Pingree, David. *Enūma Anu Enlil*, Tablets 50–51. Babylonian Planetary Omens, Part 2 (1981). Malibu, California.

Report of the Calendar Reform Committee, produced by the Council of Scientific and Industrial Research on Behalf of the Government of India (1955). New Delhi.

Robbins, Frank E. "P. Mich. 149," ed. and trsl. F.E. Robbins: *Michigan Papyri Vol. III. Papyri in the University of Michigan Collection. Miscellaneous Papyri*, ed. J.G. Winter (1936), pp 62–117. Ann Arbor.

Robertson, J. "Catalogue of 3539 Zodiacal Stars for the Equinox 1950.0," *Astronomical Papers of the American Ephemeris* 10 (1940). Washington.

Rochberg-Halton, Francesca. "Elements of the Babylonian Contribution to Hellenistic Astrology," *Journal of the American Oriental Society* 108 (1988), pp 51–62.

---. "Babylonian Horoscopes and their Sources," *Orientalia* NS 58 (1989), pp 102–123.

Rochberg, Francesca. *Babylonian Horoscopes* (1998). Philadelphia.

---. "Babylonian Horoscopy," *Ancient Astronomy and Celestial Divination*, ed. N.M. Swerdlow (1999), pp 39–59. Cambridge, Mass.

---. "A consideration of Babylonian astronomy within the historiography of science," *Studies in History and Philosophy of Science* 33 (2002), pp 616–648.

Sachs, Abraham. "A Classification of the Babylonian Astronomical Tablets of the Seleucid Period", *Journal of Cuneiform Studies* 2 (1948), pp 271–290.

---. (1952, 1): "Babylonian Horoscopes," *Journal of Cuneiform Studies* 6 (1952), pp 49–75.

---. (1952, 2): "A Late Babylonian Star Catalogue", *Journal of Cuneiform Studies* 6 (1952), pp 146–150.

---. (1974): "Babylonian Observational Astronomy", *Philosophical Transactions of the Royal Society of London* A 276 (1974), pp 43–50.

Scaliger, Joseph Justus. *Opus de emendatione temporum* (1629). Geneva.

Schnabel, Paul. *Berossos und die babylonisch-hellenistische Literatur* (1923). Leipzig.

---. "Kidenas, Hipparch und die Entdeckung der Präcession," *Zeitschrift für Assyriologie* 37 (1927), pp 1–60.

Seneca, Lucius Annaeus. *Naturalium Quaestionum*: Latin text with French trsl.: *Questions naturelles*, ed. and trsl. P. Oltramere (2 vols., 1929). Paris.

English trsl. of *Naturalium Quaestionum* III, 29 by I.P. Cory in Cory's *Ancient Fragments* (1975 repr. from 1832), p 328. Minneapolis.

Spiegelberg, W. "Ein neuer astronomischer Text auf einem demotischen Ostrakon," *Orientalistische Literatur-Zeitung* 5 (1902), cols. 6–9.

Strano, G. "The Absent Star," *Nuncius* 14 (1999), pp 19–29.

Sūrya-Siddhānta, trsl. E. Burgess (1860). New Haven.

Theon of Smyrna. *Mathematical Knowledge Useful for Reading Plato*: French trsl.: *Théon de Smyrne, philosophe platonicien. Exposition des connaissances mathématiques utiles pour la lecture de Platon*, trsl. J. Depuis (1966). Bruxelles.

Tester, Jim. *A History of Western Astrology* (1987). New York.

Toomer, G. J. "Hipparchus and Babylonian Astronomy," *A Scientific Humanist: Studies in Memory of Abraham Sachs* (Occasional Publications of the Samuel Noah Kramer Fund 9, ed. E. Leichty et. al., 1988), pp 353–362. Philadelphia.

———. Toomer (1984): see *Ptolemy's Almagest*.

Transactions of the International Astronomical Union III (1929). Cambridge.

Valentine, Peter. "Intrinsic Properties Defined," *Philosophical Studies* 88 (1997), pp 209–219.

Van der Waerden, B. L. SA: *Science Awakening* (2 vols., 1974). Leiden-New York.

———. (1952): "Das Grosse Jahr und die ewige Wiederkehr," *Hermes* 80, pp 129–155.

———. (1953): "History of the Zodiac," *Archiv für Orientforschung* 15, pp 216–230.

———. (1968): "The Date of Invention of Babylonian Planetary Theory," *Archive for History of Exact Sciences* 5 (1968), pp 70–78.

———. (1970): "Das heliozentrische System in der griechischen, persischen und indischen Astronomie," *Neujahrsblatt der Naturforschenden Gesellschaft in Zürich auf das Jahr 1970*.

———. (1974): "The Earliest Form of the Epicycle Theory," *Journal for the History of Astronomy* 5 (1974), pp 175–185.

Vogt, Heinrich. "Versuch einer Wiederherstellung von Hipparchs Fixsternverzeichnis," *Astronomische Nachrichten* 224 (1925), Nr. 5354–55, cols. 17–54.

Weidner, E. F. *Gestirn-Darstellungen auf babylonischen Tontafeln* (1967). Vienna.

———. (1919): "Babylonische Hypsomatabilder," *Orientalistische Literatur-Zeitung* 22 (1919), cols. 10–16.

———. (1924): "Ein babylonisches Kompendium der Himmelskunde," *American Journal of Semitic Languages and Literatures* 40 (1924), pp 186–208.

———. (1931): "Der Tierkreis und die Wege am Himmel," *Archiv für Orientforschung* 7 (1931–1932), pp 170–178.

Weinstock, Stefan. "Lunar Mansions and Early Calendars," *Journal of Hellenic Studies* 69 (1949), pp 48–69.

Printed in the USA
CPSIA information can be obtained
at www.ICGtesting.com
LVHW020501201124
797133LV00001B/131